LIGHT AND WATER

Capt. F. A. Bligh.]

I.

SILS BASEGLIA, UPPER ENGADINE.

A Winter Afternoon.

LIGHT AND WATER

A STUDY OF REFLEXION AND COLOUR IN RIVER, LAKE AND SEA

BY

SIR MONTAGU POLLOCK, BART.

ISBN: 978-93-5128-806-0(PB)

First Published in 1903
Indian Reprint in 2017

Published by

Kalpaz Publications
C-30, Satyawati Nagar,
Delhi – 110052
E-mail: kalpaz@hotmail.com
Website: kalpazpublications.com
Ph.: +91-9212142040

Cataloging in Publication Data—DK
Courtesy: D.K. Agencies (P) Ltd. <docinfo@dkagencies.com>

Pollock, Montagu, *Sir,* 1864-1938, author.
Light and water : a study of reflexion and colour in river, lake and sea / by Sir Montagu Pollock, Bart.
pages cm
"First published in 1903"—Title page verso.
Includes index.
ISBN 9789351288060

1. Reflections. 2. Color. 3. Water. I. Title.

NC757.P65 2017 DDC 741.2 23

PREFACE

IN presenting the following essay to the public, the author disclaims at the outset all pretension to teach anyone how to paint. His aim has been to give his readers an elementary acquaintance from the scientific side with the subject of the reflexions and colours seen in water, and he believes that an artist who will approach it from this standpoint will find a study of the questions discussed in these pages to be both of interest and service. In this, as in other cases, a true enjoyment of nature will not be marred, but rather increased, by a closer examination of her methods.

The book being intended primarily for the use of the artist or art student, there is no attempt to make the account of the subject scientifically complete, even in the ground which these pages cover. The principles involved are of a very elementary nature, and not likely to discourage the most diffident. While an endeavour has been made not to sacrifice accuracy to simplicity of treatment, all technicalities have been excluded as far as possible, though occasionally admitted as explanatory notes. The author will be satisfied if the perusal of his remarks on a topic which for some little time has afforded him food for thought and consideration, will give pleasure or profit to the practical artist and receive the approbation of those better qualified than himself to judge of the truth of his conclusions. He is fully conscious of the incompleteness of his treatment of a subject which even his limited opportunities for observation have shown to offer a far larger and more attractive study and to involve many more interesting points than might at first sight appear.

Many readers no doubt will remember that Ruskin has devoted three chapters of the first volume of "Modern Painters" to the "Truth of Water," and has enriched them with a wealth of illustration from the works of famous artists, both ancient and modern. The present writer's own pursuit of the subject, though followed along a somewhat different line, has yet only served to increase his admiration of the great teacher's marvellous insight and power of observation.

The author's best thanks are due to Mr. H. W. O. Hagreen, of Wellington College, Professor Threlfall, Dr. Aitken and others for their kind and suggestive criticisms. He is indebted for some of the photographs reproduced in this volume to Captain F. A. Bligh and Messrs. G. R. Ballance, F. Newington, H. P. Robinson, A. Spuhler, F. M. Sutcliffe and C. E.

Wanless. With the single exception of Plate XXIV, the half-tone blocks have been engraved by direct photo-process without the addition of any hand work or retouching.

CONTENTS

LIST OF PLATES

LIGHT AND WATER

INTRODUCTION

WATER, whether still or in motion, has so great an attraction for the lover of nature, that the most beautiful landscape seems scarcely complete without it. There are no effects so fascinating as those produced by the reflexions in nature's living mirror, with their delicacy of form, ever fleeting and changing, and their subtle combinations of colour. But though water owes its chief characteristics to the highly reflective power of its surface, it possesses in its transparency another attribute which distinguishes it still more from the lifeless metallic mirror, and is of the greatest importance where the question of colour is concerned. In the following pages an attempt has been made to show, on the one hand, how the various phenomena of reflexion are produced in accordance with natural laws, and, on the other, to what extent the colours we see in water are inherent in the water itself, visible in virtue of this property of transparency, or borrowed from the sky and neighbouring objects.

The true artist will always be guided by his eye rather than by any rules of science, and will instinctively seize the characteristics of water, still or in movement, and faithfully reproduce them; and some may regard with impatience any attempt to dissect these wonders of nature, or to reduce her magic to a dry system. But in his study of anatomy or perspective the artist does not disdain to call in science to his aid, and in the same way the student will find that an elementary knowledge of the fixed rules which light obeys as it falls upon water or emerges from it, will help him in no small degree to avoid errors of drawing or colouring which he might otherwise have perpetrated, and may also prove a valuable assistance to his memory when he attempts to record effects which are of a nature so essentially changing and fugitive.

In the first chapter is explained the formation of the image in perfectly still water. In the second the water is supposed to be rippled, and the consequent distortion of the image is discussed. In neither of these chapters is the colour of the water itself considered; we are so far occupied only with the form of the reflexions, and the transparency of the water is for the present ignored. In the subsequent chapters we are more particularly concerned with the question of colour. In the third chapter, the water being assumed to be still, it is shown to what extent this local colour is apparent under different conditions of vision. Finally, in the fourth chapter, the same considerations are applied to a rippled surface.

The principles involved are exceedingly simple, and when any one of them is considered alone, as in Chapters I and II, there is little or no difficulty in its application. But as a matter of fact the principles are often ignored and mistakes result. Most people will remember instances of sketches being entirely spoilt through carelessness in drawing reflexions, and some of the more common errors are pointed out in the first two chapters. These are easily corrected, but when we come to the question of colour the problem is more complicated. There are so many factors to take into account, and these factors are so variously combined in different cases, that it is difficult, and often impossible, to lay down definite directions. This is particularly the case in rough or rippled water, the colours of which are considered in Chapter IV, but even here an acquaintance with the elementary principles will be of use. It will enable the student when at work to test the correctness of his own drawings, and in his common observations will constantly quicken his perception of subtle differences of colour and tone which might easily have escaped him. To give an illustration. There are probably few people who either think that all colour seen in water belongs to the water or suppose it to be entirely due to reflexion, for no one who has attempted to paint the sea, or taken the trouble to consider the matter, can have failed to realize that, though the water shows colour of its own, it is largely affected by the colour of the sky. But many amateurs possess only an imperfect knowledge of the question, with the result that in marine paintings we frequently notice a lack of adequate harmony in colour between sea and sky, and perhaps as often an exaggerated expression of it. It is hoped that such persons will be enabled to appreciate more accurately the extent and the limits of this harmony by a consideration of the conditions on which they depend.

CHAPTER I

REFLEXIONS IN SMOOTH WATER

WE see objects by means of the light coming from them to the eye. They are either self-luminous, or, as more frequently happens, are visible by means of borrowed light; that is to say, they are illuminated by some external source, and from their surfaces a portion of the light they receive radiates in all directions. Light travels in straight lines unless some obstacle divert it from its course; and the passage of light in a straight line we call *a ray.*[1]

When a ray of light strikes a polished surface it is bent back or *reflected*, continuing its course in a new direction, and in order to understand the nature of "reflexions" in water we must be familiar with the law that every ray of light obeys when it falls upon such a surface. This law states that the angle of reflexion is equal to the angle of incidence; in other words, that the ray leaves the reflecting surface at an angle equal to that at which it falls upon it. For instance, if AB (Fig. 1) represent the surface of a plane mirror, a ray of light falling from a point P in the candle flame on to the mirror at C, will be reflected towards D, the angle PCB being equal to the angle

[1] *I.e.*, a beam of light of such minute transverse dimensions that it does not materially differ from a mere line.

DCA. Similarly the ray PF will be reflected along FG, and the ray PH along HK.[1]

The application of this exceedingly simple law affords an explanation of all the phenomena of reflexion. We have a familiar—though not very strict—analogy in the case of the reflexion of a ball from the

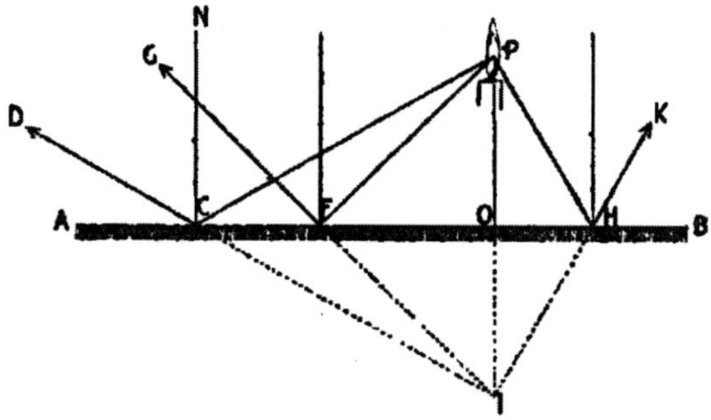

Fig. 1. The law of reflexion.

cushion of a billiard table, which may serve as an

[1] In scientific works it is found more convenient to measure the angles from the "*normal*" (or perpendicular) to the surface of the mirror at the point where reflexion takes place. In the figure the surface of the mirror is supposed to be perpendicular to the plane of the paper. If CN be drawn at right angles to AB, the angle PCN, contained between the incident ray PC and the normal CN, is called the angle of incidence, and the angle DCN, between the reflected ray CD and CN, is called the angle of reflexion. The plane PCN (*i.e.*, the plane of the paper) containing the incident ray and the normal, is called the plane of incidence. The law of reflexion, completely stated, is as follows:

The reflected ray lies in the plane of incidence, and the angle of reflexion is equal to the angle of incidence.

So the reflected ray CD lies in the same plane as CN and PC, and the angle of incidence PCN is equal to the angle of reflexion DCN, being in this case about 60°. At F the angles of incidence and reflexion are each about 45°, and at H about 30°.

illustration of the law. Referring again to Fig. 1 let us now suppose AB to represent the cushion of the table; if the player's ball is at P, and he wishes to hit another ball at D off the cushion, he must play his ball so that it strikes the cushion at C; if he wants to send his ball to G, he must play for F, and so on. In practice, owing to the imperfect elasticity of the rubber cushions (unless it have a counteracting "side" or spinning motion) the ball will travel slightly more nearly parallel to the cushion after impact than before.

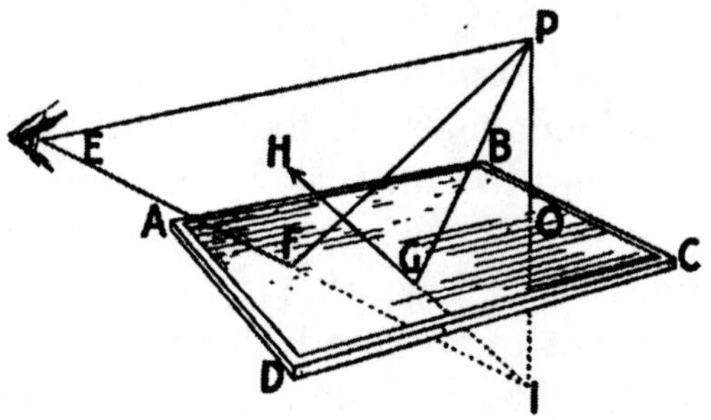

Fig. 2. Image of a point in a horizontal mirror.

In Fig. 2 we assume that there is a mere point of light at P, and that the eye is placed at E. ABCD is the perspective view of a horizontal mirror. There will be two rays of light going from P to E, (i) the direct ray PE, and (ii) the ray PFE, which reaches E after reflexion on the surface of the mirror at F; so that, in addition to the actual luminous point P, we shall see a second point of light I in the direction EF, the distance FI being equal to the distance FP. This second point of light, which appears at I, exactly as far beneath the surface of the mirror as P is above it and in the same vertical line with P, is called the *image* of P. If the eye is moved to any other point, such as H, the image still occupies the same position I; in other words, all rays from P falling upon the mirror appear after reflexion to proceed from I, the image of P. In the same way, if in Fig. 1 the lines DC, GF and KH be produced backwards, they will be found all to meet in a point I, and if a straight line be drawn from P to I, this line will be at right angles to, and bisected by, the surface of the mirror at the point O.[1] So that to find the image in a plane mirror of any point we have the following rule: *draw a perpendicular from that point to the surface*[2] *of the mirror, and produce it until its length is doubled.* Thus it is evident that I is fixed relatively to P, and as long as P and the mirror are stationary, I always occupies the same position, whatever the position of the observer. It follows that there can be only one image of a point in a plane reflecting surface; when we come to consider reflexions in rippled water, however, we shall find that this is not the case and that a single point may have a great number of images.

Having got the image of a single point, it is a simple matter to construct the image of a solid object. If we hold any object, as for instance a candlestick,

[1] The proofs of the elementary propositions of Optics made use of in this chapter, which follow from the law of reflexion given above, may be found in any text book of Physics.

[2] The surface of the mirror must be produced, if necessary, to meet the perpendicular. In the case of water, we may say, "draw a perpendicular from the point in question to the *level* of the water and produce it until its length is doubled"; for of course objects are visible by reflexion that are far beyond the actual extent of the water.

over the horizontal mirror (Fig. 3), the image of any point A on the base of the object will appear—no matter what the position of the observer—at *a*, as far below the surface of the mirror in a vertical line as the point A itself is above it (the perpendicular line A*a* being bisected by the surface of the mirror at O); in the same way the image of B will appear at *b*, that

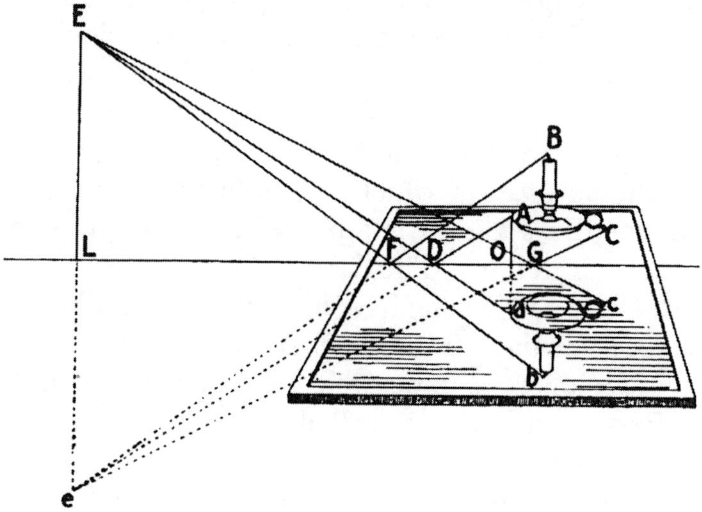

Fig. 3. The reflexion is a new view of the object reflected.

of C will appear at *c*, and so on for every other point; so that we get an inverted image *abc*, in which every part of the object is exactly reproduced. The image thus formed is called a *virtual* image, because, though it has no actual existence, the effect on the eye is exactly the same as if it were a real object. If the right hand is held up to a looking-glass a left hand appears on the other side; so we may say that the image is related to the object reflected in the same way that the left hand is related to the right. The word "inverted" in speaking of an image is therefore used in a peculiar sense. Though the image is upside down, every point in it is vertically beneath the corresponding point in the object, and when viewed so as to appear in the upright position, right has become left, and left right. Thus it is sometimes and perhaps more correctly spoken of as a "perverted" image.

Such reflexion, giving rise to the formation of an image, is called *regular*. In the case of reflexion from a rough or unpolished object, the incident light (a part only of which is reflected) is very much scattered, as if the surface on which the light falls were made up of countless minute plane surfaces facing in every possible direction. The law of reflexion still holds good, but owing to the irregularity of the surface the eye receives light from all directions and the formation of an image is impossible. It is by means of this surface-scattering, or so-called "*irregular* reflexion" of light, that all objects which are not self-luminous become visible to us. The perfectly polished mirror, on the other hand, is invisible; we cannot distinguish its surface at all, and are only made aware of its existence by the images in it of neighbouring objects.

We have shown how the image is an exact reproduction of the object. But the *view* we get of the image is a different one from our direct view of the object. In the language of perspective, the projection on to the picture-plane of the image differs from that of the object. In Fig. 3 the candlestick and its image are drawn in perspective instead of in section in order to show this difference. In the reflexion the under-surface of the candlestick is visible, whereas, looking directly at the object, it is out of sight. Anyone looking from E would get a view of the candlestick and its image very similar to that drawn in the figure. The lines ADE, BFE and CGE represent the paths of the rays from A, B and C respectively, which reach the eye after reflexion at the surface of the mirror.[1] And if the straight lines BF, AD and CG are produced downwards they will be found all to meet at a point *e*, as far beneath the glass in a vertical line as the eye is above it, so that the view of the image seen from E is the inversion of the view of the object that we should get if we were to remove the mirror and look at it from *e*. It is necessary to note this difference, because the word "reflexion" is used to denote *the observer's view*[2] *of the image*, and not the image itself. The image always remains the same, being, as we have said, the exact inversion of the object; but the reflexion, or view we get of it, varies with our position.

In considering the reflexions seen in a sheet of water we will assume in the first place that its surface is perfectly smooth and unruffled. Reflexion takes place therefore exactly as in the case of the plane mirror.[3]

[1] The line OD, produced both ways, marks the intersection of the plane of incidence of the ray AD with the surface of the mirror. For the sake of simplicity in the figure, the points B and C are chosen in this same plane, so that F and G also lie on the same line OD produced. This line bisects the vertical E*e* in L.

[2] Or, projection of the image on to the picture plane.

[3] For the present we ignore the fact that the reflective power water is different for light at different angles of incidence. This fact is, however, taken into account in the third chapter.

Suppose the whole of the objects composing the scene before us to be actually inverted beneath the level of the water, forming a solid image, and the surface of the water to be an opening through which this image can be seen; then the reflexion that we see in the water, at whatever point we may be standing, is simply our view of the inverted image from that point. Thus we get a

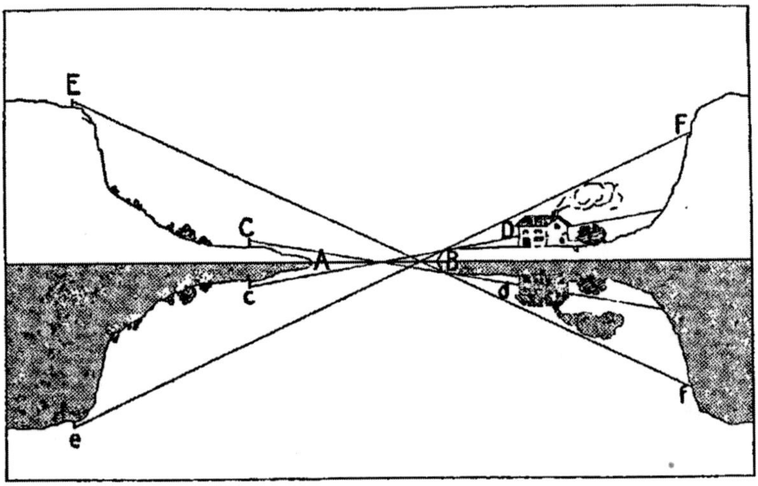

Fig. 4. The difference between the reflexion and the direct view depends upon the height of the eye above the water.

different view of the image as we change our position, the extent of our view being limited by the size of the sheet of water. This is illustrated in Fig. 4, which is intended to represent the section of a river-bed with its imaginary inverted "double," AB being the surface of the water. Standing at the point C, our view of the image of the house is limited by the further bank at B. It permits us only to see that part of the inverted house which appears to be below *d*. In other words, all of the real house which is above D (D being the corresponding point to *d*) will appear in the reflexion and no more. But if we ascend to E the image of the house is entirely hidden by the bank B, and our view of the image as a whole is limited by the line B*f*, so that we only see reflected the small portion of the cliff above the point F.

This seems to be the simplest way of regarding reflexions. But some may perhaps find the following preferable. Instead of imagining an inverted image, *let the observer suppose himself able to be lowered vertically from the point where he is standing to a position as far beneath the surface as he is actually above it. The reflexion that appears to him on the water is identical with the view he would get of the object from this imaginary point, this view being, of course, inverted.* This has already been pointed out with regard to Fig. 3, in which the view of the image from E is similar to the direct view of the candlestick from *e*. So also in Fig. 4 the reflexions seen from C and E are similar to the direct views of the actual landscape from *c* and *e* respectively, *c* and *e* being imaginary points vertically as far beneath the level of the water as C and E are above it. Again, if Fig. 7 (page 15) is turned upside down, we have the "perverted" view of the house that we should get if it were possible to look at it from a point as far below the surface of the water in a vertical line as that from which the picture is taken.

This method is the more convenient for the purposes of a diagram, as it obviates the necessity for drawing the image. For example: suppose a man standing at E (Fig. 5) to see the top of the willow,

C, the weathercock on the church tower, D, and the moon, all in one straight line. AB represents the surface of the water. To find in the reflexion, as he sees it, the height both of the weathercock and of the moon as compared with the tree C in the foreground, it is only necessary to draw the vertical EL to the level of the water at L, and produce it to *e*, so that EL is equal to L*e*. Then a straight line drawn from *e* to the weathercock will determine at what height the latter will appear in the reflexion through the branches of

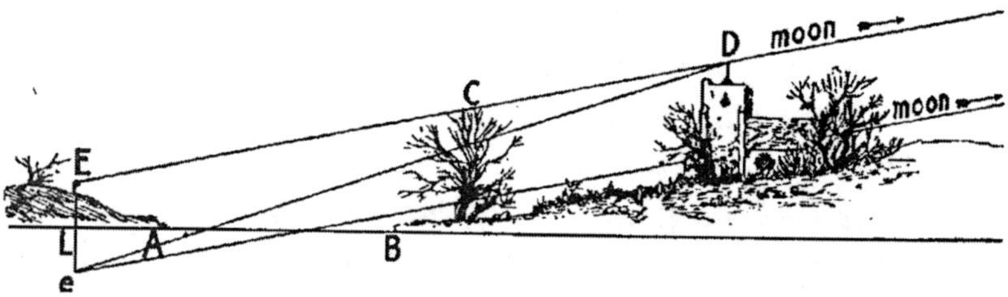

Fig. 5. The reflexion as seen from E is similar to the direct view from *e*.

the tree. This will be, as shown in the diagram, at about two-thirds of the height of the tree. The moon being practically at an infinite distance, we must, in order to find her position in the reflexion, draw a line from *e parallel to* ED. Having done this, we see that the reflected moon will be hidden behind the reflexion of the church; if the church were not in the way, the moon would appear about on a level with the top of the willow stem. The general effect will be as shown in Fig. 6.

It will now be evident to what extent the reflexions differ from the direct view of the objects reflected. The higher we are above the water the less of the landscape appears in the reflexion, whereas, on the other hand, the more obliquely we are looking on to the surface the more nearly will the actual scene and its reflexion resemble one another; in fact, if the eye were at the very level of the water, the one would be the exact double of the other. It is simply a question of the angle that the direction of vision makes with the water; if this is small the difference is small, and *vice versâ*. So it happens that, in looking down on to a lake from above, we often get nothing but sky reflexion; distant mountains, on the other hand, seen across a wide expanse of water, are reproduced in their full height (Plate II). As a picture cannot be taken from the actual level of the water, there is always some difference between the direct view and the view by reflexion, though in the case of distant objects it may be so slight as to be imperceptible. But objects in the foreground show this difference in a marked degree, those nearest to the spectator rising up in the reflexion relatively to those behind them. Thus in the photograph of the St. Moritz lake opposite the mountains seem to be exactly repeated in the water, but the roof of the little

Fig. 6. The view from E in Fig. 5.

Capt. F. A. Bligh.]

II.

ST. MORITZ LAKE.

house takes a very different position with respect to them in the reflexion from that which it occupies in the direct view. Fig. 7 affords a further instance of this difference, and here will at once be noted the change in the relative positions of the roofs and gables. In the reflexion the smaller cottage to the right rises up and half covers the tower on the barn behind,

Fig. 7.

whilst the tops of the poplars barely appear above the highest roof.

Of the two sketches that follow (Figs. 8a and 8b), the first is possible, the second impossible. In the centre of each are seen, amongst other trees, some poplars close to the opposite bank, and in the distance behind them a wooded height. These are all reflected in the still water; and it will be evident from what has been said above that in the reflexion in the lower sketch (Fig. 8b) the distant height is drawn too tall as compared with the poplars in front of it, which should overtop it *more* than in the direct view, as shown in Fig. 8a.

Fig. 8a. Correct reflexion.

As a result of this perspective rising of the nearer parts of the image, it usually happens that the reflexion

Fig. 8b. Incorrect reflexion.

of a landscape viewed across a narrow piece of still water, such as a river or canal, loses all its middle distance, so that, by reason of the different juxtaposition of the component parts of our picture, we get different effects of contrast. The tufts of grass at the water's edge, instead of being merged into the flat expanse of green meadow behind them, stand out sharply against the sky or a background of distant trees. Surfaces that are turned towards the sky, such as roofs, are foreshortened in the reflexion, whilst the reverse is the case with those that face downwards, as the under side of a boat or the inside of the arch of a bridge. The reflexion of trees and bushes at the water's edge reveals more of the dark lower surfaces of their leaves and branches than appears when we look straight at them. "We see the dark sides of leaves hanging over a stream, in their reflection, though we see the light sides above; and all objects and groups of objects are thus seen in the reflection under different lights, and in different positions with respect to each other, from those which they assume above; some which we see on the bank being entirely lost in their reflection, and others which we cannot see on the bank brought into view. Hence nature contrives never to repeat herself, and the surface of water is not a mockery, but a new view of what is above it. And this difference in what is represented, as well as the obscurity of the representation, is one of the chief sources by which the sensation of surface is kept up in the reality. The reflection is not so remarkable, it does not attract the eye in the same degree when it is entirely different from the images above, as when it mocks them and repeats them, and we feel that the space and surface have colour and character of their own, and that the bank is one thing and the water another. It is by not making this change manifest, and giving underneath a mere duplicate of what is seen above, that artists are apt to destroy the essence and substance of water, and to drop us through it."[1]

Since the image is to be regarded as a solid object beneath the water, it must (when the water is still) be so represented. To find the position of the image of any point, we have only to apply the rule, given on page 7, that any point and its image lie on the same vertical line and at equal distances above and below the level of the water. The image of a straight line is determined by the images of its extreme points, and so on. And if we have to apply the rules of perspective to the drawing of a building, we must not entirely neglect them when we come to the reflexion. The vertical lines of the image are continuations of the corresponding lines of the object, and must be drawn as such; the horizontal lines of the image are parallel to the corresponding lines of the object and must in the picture be made to converge to the same vanishing points if seen in "angular perspective," or be drawn parallel to them if seen in "parallel perspective." A further hint with regard to the perspective of reflected gables is given on the opposite page (Fig. 9)[2] and may be found useful,

[1] Modern Painters, Vol. I, Part II, Sec. V, Chap. III, § 7.

[2] The horizontal lines A, B, C, D of the object are drawn to meet in the vanishing point VP. The corresponding lines *a, b, c, d* in the image, which are parallel to them (assuming the image to have an actual existence), will therefore also meet in the same point VP. In the same manner all horizontal lines at right angles to these,

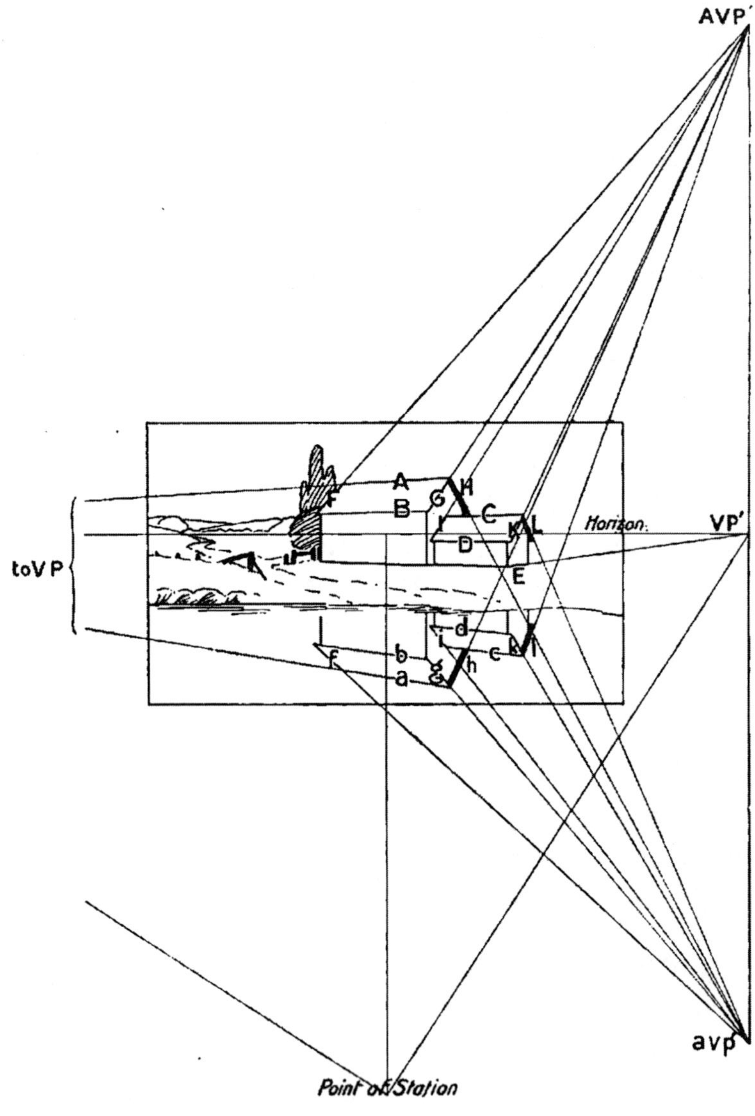

Fig. 9. The construction for drawing the gables in Fig. 7.

should great accuracy be required; though, as water is hardly ever quite still, reflexions are seldom distinct enough to need very exact drawing.

But whatever the object reflected, whether building, rock or tree, it is important to bear in mind that the image of any point lies exactly beneath that point, and must therefore be drawn vertically below it. A slight neglect of this precaution may destroy the whole force or charm of a sketch. It sometimes happens, however, that the reflexion of a sloping tree-stem appears to be shifted a little to one side, for the base of the stem, though itself in full view, cannot be seen by reflexion, its image being hidden by the image of the bank, and the first appearance of the stem in the water is therefore to one side of the base. If the water is a little rippled, so that the image is drawn out and blurred, the deception may be increased. And when the ripples are running

both in the object and the image, such as E, will converge to the other vanishing point VP′. Draw a vertical line through VP′. Then whatever pitch we choose for the roof of the house, the edges F, G, of the near side must vanish to an accidental vanishing point in this line above VP′, say AVP′. The roof of the small building has the same pitch, and therefore F, G, I and K are all representations of parallel lines and must all meet at the point AVP′. And it may easily be shown (assuming the farther and nearer slopes of the roof to be of the same pitch) that the edges H, L, of the farther side will vanish to a corresponding point, *avp′*, *as far below VP′, as AVP′ is above it.*

Now, in the image, the nearer edges *f*, *g*, *i* and *k* of the gables are parallel to the farther edges H, L in the object, and also *h*, *l* in the image are parallel to F, G, I and K in the object. Therefore the lines *f*, *g*, *i*, *k* will vanish also in *avp′*, and *h*, *l* in AVP′.

obliquely, the reflexions may sometimes be slightly inclined, as will be explained later (page 43).

But before passing to the consideration of rippled water, we may aptly conclude this division of the subject with a quotation from Ruskin, taken from his remarks on Turner's drawing of Scarborough in "The Harbours of England." "In general, throughout nature, Reflection and Repetition are *peaceful* things; that is to say, the image of any object, seen in calm water, gives us an impression of quietness, not merely because we know the water must be quiet in order to be reflective; but because the fact of the repetition of this form is lulling to us in its monotony, and associated more or less with an idea of quiet succession, or reproduction, in events or things throughout nature:—that one day should be like another day, one town the image of another town, or one history the repetition of another history, being more or less results of quietness, while dissimilarity and non-succession are also, more or less, results of interference and disquietude. And thus, though an echo actually increases the quantity of sound heard, its repetition of the notes or syllables of sound gives an idea of calmness attainable in no other way; hence the feeling of calm given to a landscape by the notes of the cuckoo."

NOTE ON THE REFLEXION OF A RAINBOW.

It is sometimes asked whether a rainbow can be reflected in still water. To this question we reply "yes" and "no." The image of a rainbow can be seen plainly enough, but it is not the image of the actual rainbow before us, but of another bow, invisible except by reflexion. As two men looking at a rainbow do not really see the same bow, so the bow seen by reflexion in water is a different one to that seen in the sky. This will be evident from the diagram opposite (Fig. 10).

Suppose a man to be standing at E, on a high cliff, with the sun at his back and rain falling on the water before him; and let EO represent the direction of the sun's rays. If, with E as apex and EO as axis, we describe a cone, of which the angle between the generating line and the axis is 42°, we get the cone FEG, or the surface on which are the rain-drops forming the bow which he sees. Imagine in the midst of the falling rain a vertical plane at right angles to his "direction of sight." The intersection of this plane with the water is shown by the line HL, and its intersection with the cone FEG is the ellipse ABCD (seen in perspective), of which the arc ABC represents the actual rainbow.

In order to find the form of the reflexion as it appears from E, we may imagine the man to be lowered in a vertical line to a point *e*, as far below the surface of the water as E is above it. Then the inversion of his view from *e* will be the actual reflexion that he sees from E (see page 12).

If the man could look from *e* at the falling rain illuminated by the sunlight, he would see a totally different bow to that which he sees from E. It will no longer be the bow ABC, but one formed on the surface of the cone *feg*, the axis of which, *eo*, is parallel to EO. The intersection of this cone with the vertical plane is the ellipse *abcd*, shown by the dotted line, of which only the portion *abc* is above the water. Therefore all the bow he could see from *e* would be *abc*, and the inversion of this, or actual reflexion as seen from E, is *akc*. For if the straight line *eb* cuts the surface of the water in M, then the ray *b*M would be reflected along ME, and the reflexion of *b*, the summit of the smaller bow (which he cannot see directly), would appear to him at *k*; and so on for any other point in the bow *abc*.

Thus we see that the reflexion consists of a smaller arc than

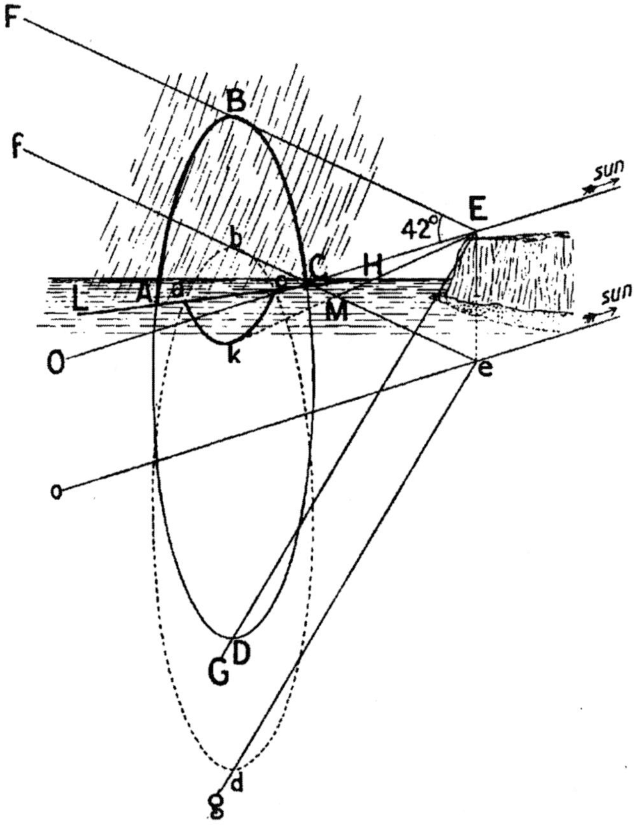

Fig. 10. Reflexion of a rainbow.

the actual rainbow, its extremities springing, not from the ends of the rainbow proper, but from points inside them. Of course, if E were at a very small height above the water, and the rain causing the bow at a great distance from E, the break between the ends of the sky bow and those of the reflected bow would be scarcely noticeable, but it seems that in a large proportion of those probably very rare cases, in which the sheet of water is large enough and smooth enough for the purpose, it might be quite conspicuous. On the only occasion on which the writer remembers to have seen the reflexion of a rainbow in water, its extremities were hidden by near objects, so that he had no opportunity of observing any break.

F. M. Sutcliffe.] [*Whitby.*

III.

SUNSET, WHITBY HARBOUR.

CHAPTER II

REFLEXIONS IN RIPPLED WATER

THE absolutely smooth surface thus far assumed, giving a perfect, but lifeless, reflexion of neighbouring objects, is not often met with except in small ponds or in sheltered corners. A sheet of water of any extent is almost always in more or less motion, the amount of which varies with the strength of the breeze. It is the gentler motions that give rise to the most beautiful and characteristic reflexions; and for the consideration of these we will now suppose there to be a slight disturbance of the water, strong enough to ripple its surface, but not so strong as to destroy all form of the image.

The following simple experiment will serve to illustrate what occurs under the new conditions. Let a small looking-glass be laid flat on the table, so that in it is seen a reflexion of the wall opposite. If the further edge of the glass be now raised, so that it is tilted a little towards one, the line of vision will also be tilted upwards, and consequently a higher part of the wall than before, or perhaps the ceiling, will be seen reflected.[1] In Fig. 11a the mirror is shown in

[1] In Chapter I we have spoken of light travelling in straight lines and of its passage from a source of light or an illuminated object to the eye. We now find it convenient to trace the rays of light in

Fig. 11a. Showing direction of lines of vision after reflexion in the horizontal mirrors A, B, C and D.

Fig. 11b. The mirrors tilted towards the observer.

four different positions, lying horizontally on a table. There is a picture FG hanging on the wall to the left. If the eye is placed at E, we see reflected at A the base, at B the middle, and at C the top of the picture. At D we shall see only the bare wall above.

Fig. 11c. The mirrors tilted away from the observer.

In other words, the reflexion of the picture extends

the opposite direction, *i.e.*, from the eye to the object instead of from the object to the eye.

We have shown how a ray of light is reflected at a plane surface in obedience to a certain law. For instance, in Fig. 2, page 6, the ray PF is reflected at F, so that it reaches the eye at E; that is to say that if we look from E towards F we shall see by reflexion the point of light P. So that, if we call EF our "*line of vision*," we may say that the line of vision is reflected at the point F on the surface of the mirror and proceeds after reflexion, in obedience to the law already stated, in the direction FP. By adopting this convention, the discussion is much simplified.

from A to C, and no farther. The lines of vision in this diagram take what we may call their *normal* directions, *i.e.*, those which they would follow after reflexion in a horizontal surface, as that of still water. Now if the mirror be slightly tipped up, as in Fig. 11b,[1] the lines of vision will all be shifted upwards from their normal directions; the reflexion of the top of the picture appears no longer in its normal position at C, but has retired to A, whilst at B and C we see the wall above, and at D the ceiling.[2] (In order to see the bottom of the picture reflected in the tilted mirror, it would be necessary to move the table forward and place the mirror at K.) So, in the case of a succession of waves coming towards or receding from us, it is evident that, as we look upon their near sides, we shall frequently see nothing but the sky reflected, whereas, had the water been smooth, we should in the same direction have seen the reflexions of objects on the opposite shore.

Fig. 12 is intended to illustrate roughly the way in which this effect is produced. Suppose that a man

[1] Here represented in each position as tilted through an angle of 5°.

[2] The effect is the more noticeable owing to the fact that if a mirror on which a ray falls is rotated (on an axis at right angles to the plane of incidence) through any angle, the reflected ray moves through *twice* that angle. For if the mirror be turned through any angle θ, the normal to it is turned through the same angle. Hence the angle between the ray and the normal is increased or diminished by θ, and therefore that between the incident and reflected rays (which is double of this) is altered by twice θ.

In the same way, if one of the mirrors in Fig. 11a is tilted through a given angle, the line of vision undergoes deflexion through twice that angle.

is looking from E at the surface of the near wave at A. In order to find the direction of his line of vision after reflexion, we may suppose there to be a tiny mirror floating on the surface of the water at A; this mirror will be tipped up towards him, and his line of vision will be along AB (the angle of reflexion being equal to the angle of incidence, as explained in the last chapter), whereas, if the water had been smooth, he would have been looking on the level surface at C, and his line of vision would have taken the direc-

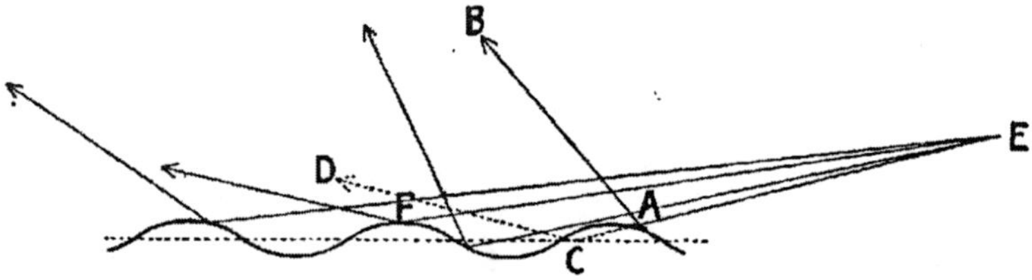

Fig. 12. Lines of vision, on striking the near sides of the waves, deflected towards the sky.

tion CD. (The dotted lines represent the condition in smooth water.) There are of course points on the surface even when the water is disturbed, as, for instance, on the crest of a wave at F, where a floating mirror would assume a horizontal position, but it is evident from the figure that in the great majority of cases the line of vision strikes a surface inclined *towards* the observer, and is therefore reflected up towards the sky, in some cases less, in others more. The result is that it is only at certain comparatively minute spots that we catch sight of the image that would appear in smooth water, these spots being surrounded by surfaces reflecting the sky so that unless the ripple is very gentle, the image is soon lost. The position of the spectator in the figure (Fig. 12) is so chosen that he can see little but the near sides and the crests of the waves, and indeed this is generally the case when one is standing close to a sheet of water and looking at a gently rippled surface at a little distance. The line of vision is so nearly parallel to the level of the water that the far sides of the waves are out of sight. The objects on the opposite shore are seen reflected in the crests of the waves (supposing always that they would be visible in that part of the water, if smooth) and between these glimpses of the image the sky appears by reflexion on the tilted surfaces. If the water is much agitated, or if one is looking in a virtually horizontal direction, as at distant waves from a position near the level of the lake, one sees nothing but their near sides, and therefore in all probability only reflected sky light.[1]

[1] In any wave-section (as in Figs. 12 or 13) there is of course only one point on the crest of each wave—*i.e.*, the summit—where the normal is vertical, and where a ray would therefore be reflected as on a horizontal plane. But on either side of this point there is a short piece of the curve where the normals are nearly vertical, and which would reflect rays to the eye from points close above or below the point which is seen reflected in the summit. Thus a conspicuous white object, such as the side of a house, gives *patches* or *strips* of white light on the crests of the waves.

It is perhaps unnecessary to warn the reader that if he could see a definite image of any object in the curved side of a wave, it would be a very distorted one (and in a concave surface an erect, instead of an inverted, image). As a matter of fact, this is hardly ever possible, and moreover does not at all concern us in the present discussion, as we are considering the combined effect of the re-

IV.

RIVER INVER.

The reflexions of the hills, drawn out by the gentle motion of the water into vertical streaks, will soon be obliterated by the stronger ripple that is rapidly spreading over the surface.

V.

SCOURIE.

Showing the disappearance in rippled water of the horizontal lines of the objects reflected. In the reflexion the top of the garden wall is vague and blurred, whilst the upright line to the left remains perfectly distinct.

Thus we have the well-known effect of a ripple spreading over the water. The image of the hills opposite, a few moments ago so vivid, becomes less and less distinct, being interrupted by minute patches of sky reflexion, while, as the breeze gains in strength, it disappears altogether, and the water gleams with light from sky or cloud. (See Plate IV.) For the reason just given all definite reflexions in water in the far distance vanish entirely with the commencement of the ripple.

So far it has been assumed that only the near sides of the waves are in view; but when the breeze is so gentle that the height of the waves is small compared with their length[1] and one is not standing too low, a certain amount of their *farther* sides, as in Fig. 13 at *a*, *b*, *c*, *d*, *e*, *f*, is also seen.

In Fig. 11c (page 27) is shown the effect of tilting the mirror *away from* us, instead of towards us, as before; the lines of vision are now deflected *below*

flexions in hundreds of minute waves, rather than a single image in any one. For our purpose it is enough to know that light from a given object reaches the eye after reflexion at a certain part of the wave.

The *circular* curves (so drawn for greater convenience) representing the wave forms in this chapter, are only rough approximations to the theoretical wave-form, which is a so-called "*harmonic curve*"

or "*curve of sines.*" It will be noticed that the harmonic curve is much flatter on the sides of the waves than the circular curve.

[1] The "length" of a wave is the distance from one crest to the next.

their normal directions; the reflexion of the upper edge of the picture has advanced to D, that of its lower edge to C, and at A and B are seen parts of the wall beneath.

Now when we look at rippled water, we see surfaces tilted at many and ever-varying angles, so that a mirror floating at any point (as supposed in Fig. 12)

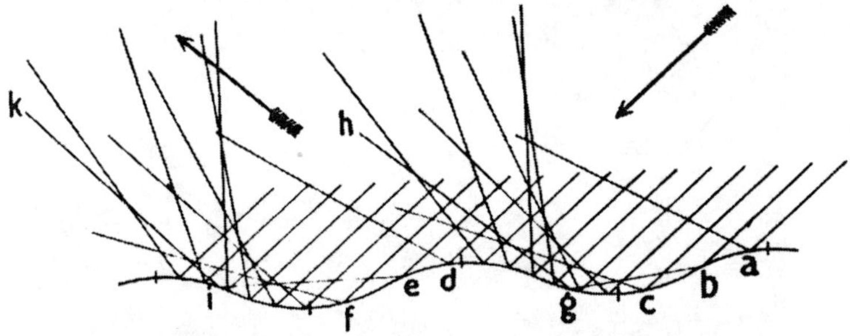

Fig. 13. Here the observer is supposed to be looking from a considerable distance on to the surface of the rippled water (in the direction of the right-hand arrow, *i.e.* at an angle of about 40° with the horizon) and the lines of vision may therefore be represented by *parallel* straight lines. If the water were still these lines would also be parallel after reflexion, taking the direction shown by the arrow on the left. The crests and troughs of the waves are marked by short vertical lines. The figure shows how few, comparatively of the lines of vision (even at this abrupt angle of incidence) strike the far sides of the waves, and how some of those that do so, as at *h* and *e*, undergo a second reflexion from the near side of the next wave. Thus there are very few lines inclined *below* the normal direction of reflexion (as represented by the left-hand arrow) compared with those tilted up towards the sky.

would be perpetually changing its inclination, swaying to and fro as the waves passed by; and therefore if we imagine the mirrors in Figs. 11 to oscillate between their positions in the second and third diagrams (though not necessarily in unison) we get an idea of

A. Spuhler.] [*Neuchâtel.*

VI.

ST. MORITZ LAKE.

The gentle movement of the water in the distance elongates and exaggerates the upright lines of the buildings. In the foreground the individual ripples become visible, breaking up the reflexion of the mountains horizontally.

what takes place in the water. By comparing these two figures (11b and 11c) we see that the position of F's reflexion moves backwards and forwards between A and D, and that of G between K and C. In other words, if we could substitute a sheet of rippled water for our table and mirrors[1] the reflexion of the picture would be prolonged towards us as far as D, and away from us as far as K. The reflexion of the wall below would blend with this as far as C, and every part of the surface nearer than A would also show light from the wall above or the ceiling.

By prolonging the lines EA in Fig. 11b and ED in Fig. 11c to *f*, making A*f* equal to AF and D*f* equal to DF, we get the position of the image of the point F in each diagram. These are the two extreme positions of the image of F, and between these it may be seen at any intervening position (its normal position being shown in Fig. 11a) so that, instead of appearing as a point, it seems to be drawn out into a vertical line.[2] Of course, if the water were more strongly rippled, the position of F's reflexion would lie anywhere between a point still farther back than A and

[1] We assume that the imaginary sheet of water is rippled by waves (travelling at right angles to the direction of vision) to such a degree that the greatest inclination of the tangent to the surface at any point is that of the mirrors in Figs. 11b and 11c, an angle of 5° to the horizon.

[2] These points, *f*, in each diagram all lie, of course, in the plane of reflexion (or incidence), that is, the plane containing all the lines of vision both before and after reflexion. For the sake of greater clearness, the intersection of this plane with the ceiling, wall and floor is shown by a broken line. In 11c, *f* lies in front of the plane of the wall; in 11b, behind it.

one nearer than D, so that the image would be still more drawn out.

We are now able to understand the simple method by which Nature produces this most beautiful effect. As the plainest possible instance, we can imagine the picture FG to represent the white side of a house, the wall below to represent the grassy bank on which it stands, and the wall above the blue sky. The reflexion of the house is drawn out to twice its natural height, the green light from the bank mingles with this for some distance, and the whole is interrupted by patches of blue sky. The motion of the water gives life to the image, increasing thereby its extreme delicacy both of form and colour. We have supposed the ripples to be parallel to the bank on which we are standing, and, indeed, this is usually the case if the water is not too deep close to its edge, for the waves, whatever their former direction, on reaching shallow water veer round and range themselves parallel to the shore line—so that these interruptions of one part of the image by another take the form of horizontal strokes or lines. This is clearly seen in the foreground of Plate VI; note also the reflexion of the headland in Plate XXXI, page 66.

Fig. 14 shows how, looking from Q, a luminous point P may be seen reflected at different points on the surfaces of succeeding waves. A number of lines, representing rays of light, are drawn proceeding from P and hitting the surface of the water, but only those are shown which after reflexion reach the eye at Q. If the water were still, P would be seen by reflexion on the smooth surface at the point X, its image

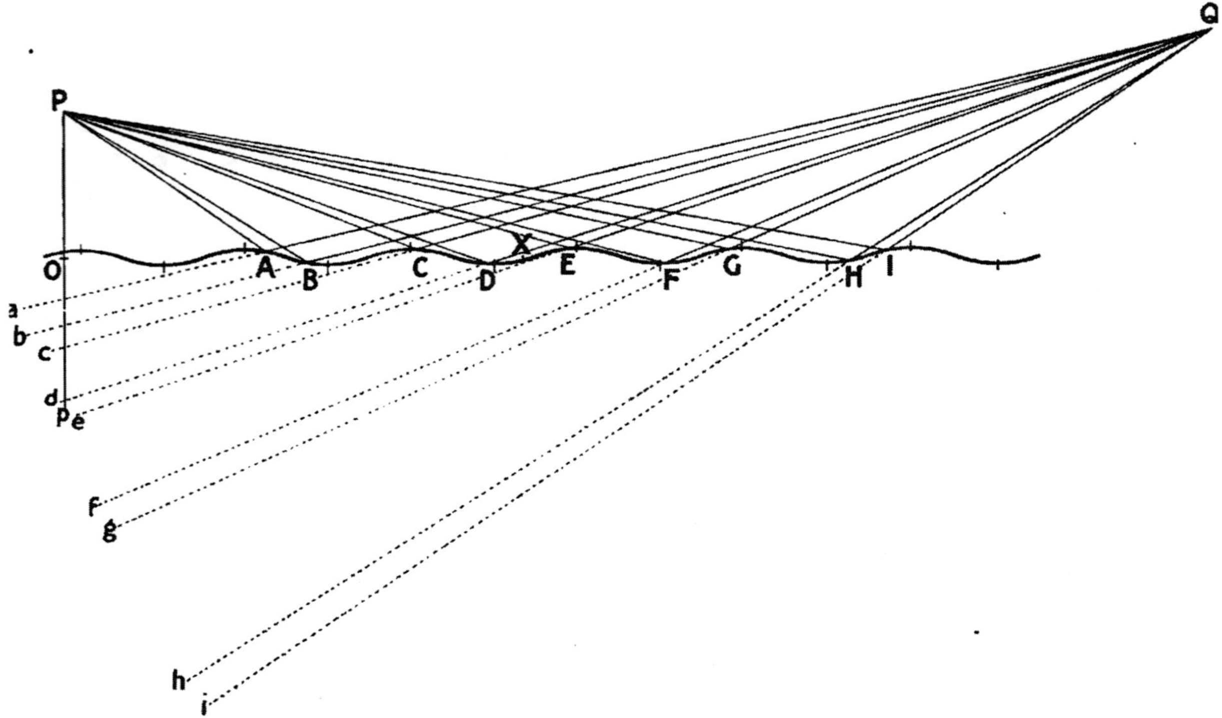

Fig. 14. The many images (a, b, c, d, e, etc.) of the point P visible from Q in the rippled water.

appearing at *p*, as far below the level of the water, O, as P is above it. Beyond X, at A, B, C, and D, P is seen by reflexion on the *near* sides of the waves[1] and the image is *lifted* above *p*, appearing at *a*, *b*, *c*, and *d*, respectively; whilst at all points nearer than X, as E, F, G, etc., it is seen by reflexion on the *farther* sides of the waves, and the image is *lowered* to *e*, *f*, *g*, etc. These points *a*, *b*, *c*, *d*, *p*, *e*, *f*, *g*, etc., all lie in the vertical plane containing P and Q (*i.e.* the plane of the paper) and when viewed from Q will therefore all appear to lie in the vertical line through P. Fig. 15 is intended to represent this view of the waves in Fig. 14 as seen from Q, with the different images of P, namely, *a*, *b*, *c*, *d*, *e*, etc., formed in them. The farther sides of the waves are indicated by deeper shading. (If the water were still, the image of P would appear at *p*.) As the waves advance towards one, the images follow one another downwards in the vertical line (or upwards, if the waves are going from one). *Thus in the rippled water the combined effect of all the images of the luminous point P, formed by reflexion*

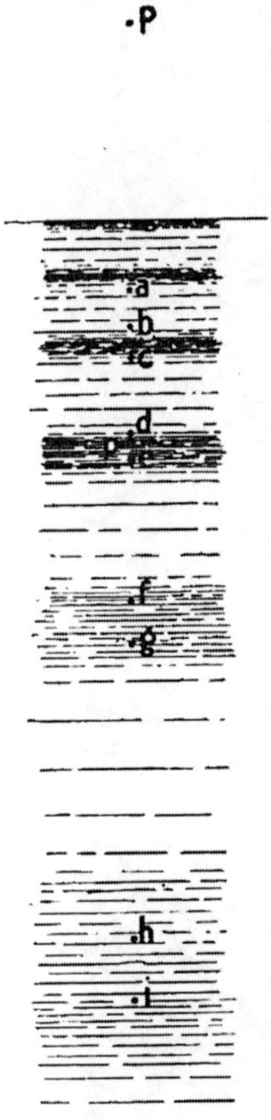

Fig. 15. P and its images, as seen from Q in Fig. 14.

[1] The crests and troughs of the waves are marked by short vertical lines.

VII.

Wide sun streak; sun high and water rough.

VIII.

A portion of a wide sun streak, caused by sunlight striking the water through a rift in the clouds.

in countless wavelets, is a vertical line of light. (See Note at end of chapter.)

This phenomenon is seen to perfection in any harbour at night, where the lights of the vessels at anchor are represented in the water by long quivering lines. When the motion of the water is very gentle, these lines are wonderfully well defined, but with the slightest breeze they become blurred and indistinct. In the reflexion of the moon, when the ripples are very regular, the image is a path of light of about the same width as the moon itself, consisting of a succession of horizontal lines, as shown in Plate III, but there are points outside this limit that occasionally catch the light, and as the surface gets more ruffled, the streak of light widens and becomes less clearly defined, so that looking from a distance the state of the water can sometimes be roughly judged by the width of this streak of sun or moonlight. When, as so often happens, the silver pathway beneath the moon widens out towards the horizon, we have an indication that there is more breeze or rougher water at some distance from the shore. Or only a remote part of the water may catch the glitter, the intervening surface being too quiet to show it at all. It is a common thing on the West coast of Scotland to see the sun reflected towards evening in the sheltered waters at the head of a loch, its image swinging placidly at one's feet, whilst in the more open sea beyond there is a patch of dazzling light under the sun, reflected from the near sides of the waves.

But the width of this streak of light also depends

very largely upon the *height* of the sun or moon above the horizon. As the sun mounts in the sky it gets wider and vaguer, disappearing altogether towards midday if the sea is very smooth; in the afternoon it narrows again, becoming more brilliant up to a certain point of blinding intensity, and more and more clearly defined as the sun descends to within a few diameters of the horizon, when it vanishes rather suddenly. In a choppy sea, with the sun at its highest, sparkles of directly reflected sunlight reach the eye from all parts within a wide angle.

Plates VII and VIII are photographs of wide "sun-streaks," the sun being high and the sea fairly rough. In Plate VII a stiff breeze is blowing, and, at any rate in the distance, the streak extends to the width of the photograph. In the lower view the space, which, if the sky were clear, as in Plate VII, would be filled with dazzling light, is almost entirely shadowed by clouds, which allow the direct sun-rays to reach the water through a narrow opening only. The sun is at about the same height as in Plate VII, but there is less breeze, so that the width of this luminous space is less than in the upper view. In Plates IX and X the sun is much lower; but here again a stiff breeze is blowing, so that the water is much ruffled and the streak is still fairly wide in comparison with the size of the sun. Plate X was taken about forty minutes later than Plate IX, the breeze blowing, as far as could be judged, steadily all the time, and shows the narrowing of the streak as the sun sinks. We have already seen in Plate III a specimen of the sun's reflexion in a very light ripple;

IX.

X.

Showing the narrowing of the streak of light as the sun declines.

Plate XI shows the streak just before its disappearance and about as clearly defined as is possible in the open sea.

In considering the formation of such a streak of light, we started by assuming the spectator to be looking *straight across* a succession of very regular waves, which, it was then shown, would cause a luminous point to appear in the water as a vertical line. But it might be argued that very regular waves travelling *at right angles to* the direction of sight would give a *horizontal* line in the water, and that therefore, if the sea were rough, with the sides of the waves facing in all directions, we should get, as the reflexion of a luminous point, no streak at all, but only a mere patch of light.[1] That these cross waves are, however, of little account as compared with those

[1] The statement on page 36, that the points *a*, *b*, *c*, *d*, etc., in Fig. 14, all lie in the plane of the paper, is based on the supposition, made for the sake of argument, that the waves are perfectly regular, with the lines of their crests at right angles to the plane of the paper. But as a matter of fact, the waves are often very far from regular, and the motion of the water is generally exceedingly complicated, different wave-systems crossing each other in all directions, so that the normals to the surface are inclined in every possible direction within a certain angle from the vertical. The point P would under these circumstances be seen reflected at points on the surface of the water to the right and left of the vertical plane containing P and Q, and the images thus formed will lie to the right and left of the vertical through P (Fig. 15), so that, as we have said, the combined effect of all the images will be a vague *streak* rather than a well-defined line. In order to meet the objection stated above, viz., that unless the movement of the waves towards or from the observer were much more marked than in other directions, the vertical streak would entirely disappear, let us consider the opposite extreme and suppose the waves to be moving across

travelling to or from the spectator, may be illustrated by a rough experiment, such as the following.

A sheet of glass is bedaubed on one side with vaseline or other grease, and then carefully drawn,

from right to left or *vice versâ*, at right angles to their former direction.

Suppose the curve in Fig. 16 to represent the section of the waves through X (Fig. 14) now running in the transverse direction, but of the same height and length as before. Q and P occupy the same positions, and are here shown in elevation, the vertical line PX representing the plane of the paper in Fig. 14. The normal to the surface of the water at any point would reach this plane at N or thereabouts (the position of N being deduced from Fig. 14). By drawing from N a line NK at the greatest inclination that the normal can assume (in this case 23° with the vertical), we get the point K as the farthest possible point to the right of PX at which P can be seen reflected. Figs. 14, 15 and 16 all being drawn to the same scale, we now see how small is the lateral shifting of the image in comparison with the vertical displacement. In choppy water the normals to the surface point in all directions within a certain limit, and the line of light becomes, as we have said, widened into a streak, but the greater the distance from the central line, the less the number of points that catch the light, so that the streak has no definite edges and fades very gradually away.

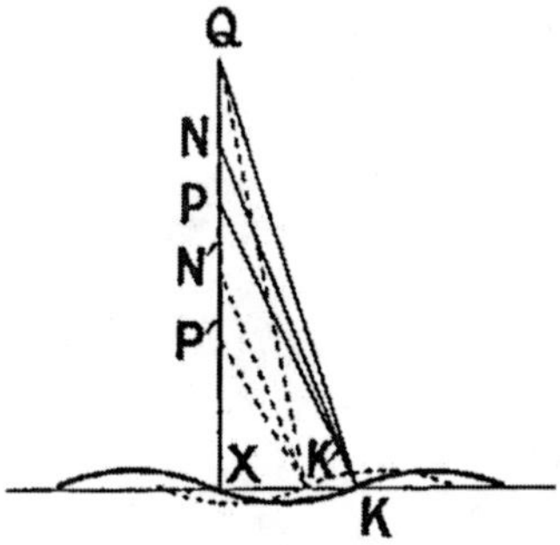

Fig. 16.

If, in Fig. 16, we suppose the luminous point to be lowered to P′, we get K′N′ for the normal, and the position of K′ now shows the limit of possible reflexion of P′. Thus, as the position of the object (or the observer) is lowered, the streak narrows. This we have already seen to be the case with the path of light below the sun (Plates IX and X).

greased side down, over a cloth, with the result that, on lifting it, its surface is smeared in parallel straight lines. The glass is then placed, greased side upwards, so that the light from a candle is reflected from it to the eye at an oblique angle; and, first, with the lines of grease *across* the direction ofvision. A long, well-defined streak will be seen to run vertically through the image of the candle flame thus (Fig. 17a). Now let the glass be turned round. The streak will gradually become inclined, but very slowly, so that when the glass has been turned through half a right angle, the streak will only have reached the position shown in Fig. 17b. When, however, the glass has been turned through a whole right angle, so that the lines of grease lie *along* the direction of vision, there will remain only a very short horizontal streak of light through the image, as in Fig. 17c.

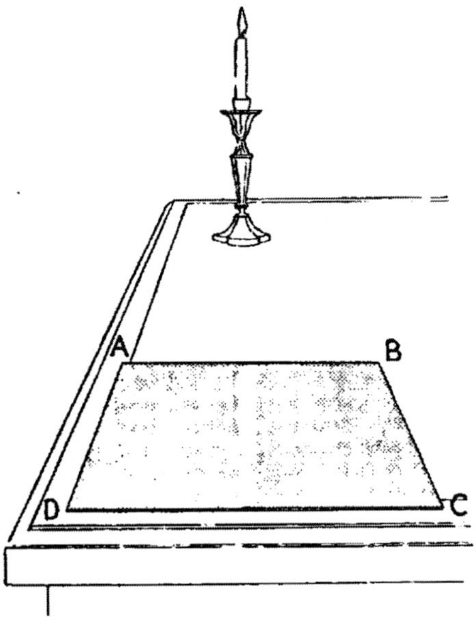

Fig. 17a. Reflexion of a candle flame in greased glass (lines of grease parallel to AB).

We have here a rude imitation of the effect produced by reflexion from very regular ripples. In each figure the lines of grease run parallel to AB and CD.

Fig. 17a may therefore be taken to represent the general effect of waves travelling directly towards or away from the spectator, and here the streak is very clearly defined and far too long to appear in its entirety on the greased glass. In Fig. 17c, which illustrates the condition when the waves are running at right angles to the direction of sight, the streak is very short and ill-defined. Thus we see that when, as is generally the case, one is looking obliquely at the surface of the water, the action of the latter kind of waves may almost be neglected in comparison with that of the former. A combination of the two forms of streak, that is to say, the length of the vertical streak with the width of the horizontal one, would give approximately the effect produced if the wave-surfaces faced equally in all directions.

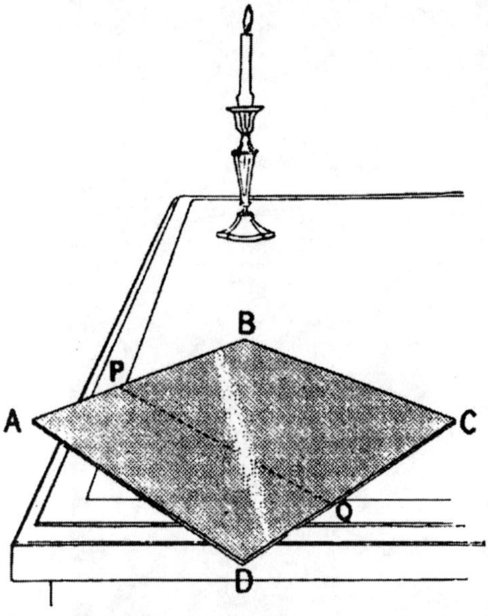

Fig. 17b. Glass turned round horizontally through 45°.

When the photographs shown in Plates IX and X were taken, the wind happened to be blowing hard from the north, and the ripples were travelling, as may plainly be seen, *across* the direction of sight.

F. M. Sutcliffe.] [Whitby

XI.

Narrow sun streak; sun low and sea calm.

Yet this fact did not prevent the formation of *vertical* streaks. If the strongly illuminated portions of the little waves are closely examined in the original photograph, they are found each to consist of a number of minute spots of light, which no doubt represent surfaces turned towards the sun.

The experiment described above also shows how, when the ripples are very regular and travelling *obliquely* to the direction of sight, the streaks formed by reflexion, instead of being vertical, may slope a little to one side or the other. But it will be noticed that the streak in Fig. 17b is only slightly inclined, and is far from having reached a position (viz., PQ) at right angles to the lines of grease. In nature this effect, except on the surfaces of the nearer waves, is so unusual and inconsiderable, that in a picture it might give an appearance of careless drawing. In the foreground, however, we sometimes get strange effects of distortion or bending of the image, as, for instance, in the smooth waves in the wash of a boat moving through calm water.

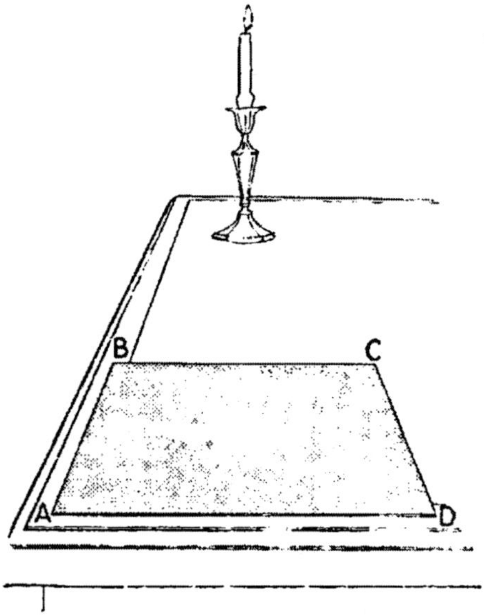

Fig. 17c. Glass turned through a right angle from its first position.

We have seen how a point is elongated into a vertical streak, a lamp reflected in gently moving water appearing as an upright line of light; and so it is that in these drawn-out reflexions in rippled water the perpendicular lines of the object are retained whilst its horizontal lines disappear. In the reflexion of a house having rows of windows evenly arranged one above another, it will be noticed that it is often impossible to distinguish the windows of the different floors, they having become merged together into vertical lines extending from the top to the bottom of the house, as can be seen in Plate VI. Again, in the reflexion of a railing in rippled water the upright posts may be perfectly distinct, though the rails connecting them have entirely disappeared, and, in the same way, the piers of a bridge will remain long after its arches have vanished.

In a stretch of water of any size there is almost always enough motion to destroy the horizontal lines of the image, so that, in drawing reflexions, this most characteristic feature must not be forgotten. The vertical edge of a white cottage wall is sharply defined, but the horizontal lines of its red-tiled roof have escaped us, and in the water white dissolves into red and red fades away into the green of the wood behind. (See Plate V, page 30.) The trunks of the trees and the masts of boats stand out prominently, whilst other parts of the reflexion picture are blurred and indistinct or vanish altogether (Plate XII); the very upright and straight stems being more distinct even than those that slope only a little. A beautiful example of this effect may be seen in the reflexion of a

XII.

ON THE BURE.

Showing the persistence in rippled water of the upright lines of the object. Note the disappearance in the reflexion of the sloping bough on the big tree.

XIII.

A NORFOLK FARM.

A gentle ripple at the opposite end of the pond throws the reflexions

floating swan in gently rippled water; the upright neck is represented by a long and well-marked vertical streak, whilst the body of the swan is lost among the ripples. Plate XIII, showing the action of a light breeze on the farther side of the pond, is a good instance of this tendency of rippled water to emphasize the vertical lines.

These "streaky" reflexions, so characteristic of moving water, are perhaps seen at their best when the ripples are large[1] and unruffled by the slightest breeze, and the water assumes that oily appearance so often seen at sea on a perfectly calm day. The rocky shore will now cast no definite image, its reflexion being apparently made up of innumerable upright streaks of every shade, a dark shadow or stain on the rocks giving rise to a dark streak in the water, and a white stone or other light object yielding a light streak. Each of these streaks is composed of a succession of short horizontal lines or may consist of one continuous descending zigzag. On a small sheet of water the ripples are necessarily small also, and though in that case the general effect may be much the same, the greater charm is undoubtedly given by the way in which the larger ripples break up the vertical streaks of which the reflexions are composed into patches or waving lines of colour—an effect which lends itself to bold treatment on the canvas. This familiar effect is, however, not easy to imitate correctly

[1] The *large* ripples here assumed are due to some distant disturbance in the water. Ripples caused by the immediate action of the wind are far smaller than these and their sides are steeper, so that the image is nearly obliterated.

in painting without close observation of the actual scene. Nature seems at first sight somewhat partial in her decision as to which object shall cause a streak, and which, though conspicuous enough on the hill side, shall have no visible counterpart in the water. We shall find on looking more intently that objects low down close to the water's edge are not reflected at all,[1] and the same may be said of those above a certain height, whilst between these limits a prominent piece of colour is represented by a well-marked streak starting from the very edge of the water. But sometimes a bright patch on the rocks occurs just above or below an equally dark patch, so that the one kills the other, and no trace of either appears in the water. On the other hand, two or more scarcely noticeable spots or patches, which happen to lie in the same vertical, combine unexpectedly to yield a definite streak.[2] Any upright crack or stain on the rocks is of course exaggerated in the reflexion whilst the horizontal ones are lost. Should a breeze spring up, ruffling the surface at first gently, the reflexion of the rocks gradually dissolves away, until the origin of the broader and less distinct streaks that now remain may be traced to the outline of the cliffs against the sky.[3] Wherever the cliff is highest the water beneath

[1] With the very gentlest tremble of the water the phenomenon is reproduced on a small scale, every stone at the water's edge then giving a streak.

[2] Thus it sometimes happens that in the reflexion of a featureless hillside we get a series of very evident streaks, the origin of which is by no means so evident.

[3] The reflexion of a headland standing out to sea often terminates

G. R. Ballance.] [*St. Moritz.*

XIV.

DAVOS LAKE.

Delicate cloud streaks in very gently rippling water.

is darkest, and *vice versâ*. A still greater disturbance will leave only vague streaks from dark and light spaces in the sky above. For even when water is much ruffled there generally remain—long after all definite image is lost—broad bands of light on the surface radiating from the eye towards those parts of the sky which are most luminous. (See Fig. 19, page 58.) This is an important principle in marine painting. The sea, however rough, hardly ever shows the same tone or colour in all directions. Within the limits of a picture it generally varies somewhat, though that part of the sky which gives rise to the difference of tone may be far above the usual limits of the field of vision. By watching on sunny days the display of light under the sun—for at midday it is often too wide to be called a "streak" of light—we get a hint as to the way in which the water is affected by different regions of the sky. We saw on page 38 that with a fair movement of the water this display persists from the time of the sun's position at a high altitude—no definite limit can of course be named—until it is only a few degrees above the horizon, thus showing how great a depth of sky affects every part of the rippled water. But we found that the width of the dazzling surface beneath the sun depended not only upon the sun's altitude but also upon the state of the water, that, as the sun approached the horizon, the golden pathway narrowed and finally disappeared, though lasting longer in smooth than in rough water. So, in a picture, it would be erroneous to make the rippled

abruptly with the top of the bluff and takes no account of the sloping base or talus below.

surface of the sea affected in tone by a bank of cloud lying low down on the horizon; or again, to draw *well-defined* streaks as due to very high[1] clouds, for we have seen that the upper parts of the sky affect large widths of sea. Above the limit at which the sun streak vanishes, the lower the clouds the better defined are the streaks which they cast. The latter are often most fascinating to watch in their rapid changes, as when thrown by brilliant white clouds floating in a blue sky on to water rippled by a light and shifting breeze. As in the instance of the rock reflexions described above, we may again be puzzled for a moment by the presence or absence of a streak in an unexpected quarter. Two or three insignificant clouds combine to form an unmistakable streak, whilst a bright conspicuous mass—too high, or too low, or possibly neutralized by some dark mass above or below it—casts none. It is curious to see how an almost horizontal layer of clouds only slightly thicker or darker in places throws definite streaks across the nearly smooth water, so determined is nature to convert horizontal into vertical. "If we see on an extent of lightly swelling water surface the image of a bank of white clouds, with masses of higher accumulation at intervals, the water will not usually reflect the whole bank in an elongated form, but it will commonly take the eminent parts, and reflect them in long straight columns of defined breadth, and miss

[1] By "*high* clouds" are meant those at a great angular elevation above the horizon. A streak cast by such a high cloud would not reach as far as the horizon, for a streak could hardly extend *beyond* the object causing it.

XV.

A HIGHLAND LOCH.

Well-marked cloud streak.

XVI.

LOCH GLENCOUL.

Cloud reflexion in rougher water.

the intermediate lower parts altogether; and even in doing this it will be capricious, for it will take one eminence, and miss another, with no apparent reason; and often when the sky is covered with white clouds, some of those clouds will cast long tower-like reflections, and others none, so arbitrarily that the spectator is often puzzled to find out which are the accepted and which the refused."[1] With a stiff breeze all, except only the very broadest of these cloud reflexions vanish, and, if the upper sky is overcast, the water becomes a cold gray, darkening to an almost inky colour.

Three views of clouds and their reflexions are given. A photograph generally fails to record the more delicate differences of tone in water subjects, and something more is inevitably lost in the process of reproduction. Plate XIV, however, gives fairly well the general appearance of cloud reflexions in a very gentle ripple, and shows well-defined vertical streaks, due to the disposition of the clouds in the lower half of the sky. Plate XV is a photograph of a cloud-streak in a stronger ripple, whilst XVI gives a still broader effect of the same kind in roughish water. In this latter case the extreme darkness of the water to the left is partly caused by a heavy cloud too high to appear in the photograph.

Enough has now been said to illustrate the tendency of moving water to resolve all reflexions into what, for want of a better term, we have called vertical "streaks."[2] Plates XVII, XVIII and XIX are

[1] "Modern Painters," Vol. I, Part II, Sec. V, Chap. I, § 12.

[2] See Plates IV, XIII and XXIII.

given as further instances of this general tendency. In the first of these the ripple (except in the wash of the steamer from which the view was taken) is very slight. The vertical lines of the buildings are distinctly repeated in the reflexion, the horizontal lines having, of course, disappeared. In Plate XVIII there is more ripple and a greater amount of sky reflexion, leaving comparatively vague streaks from the domes and towers of the church. Plate XIX represents a further stage. The motion of the water is much livelier, and nothing approaching to a definite reflexion of the yacht appears—not a sign of the mast, and only the vaguest of streaks from the sails. We have added a photograph of reflexions in slightly rough ice (Plate XX), which may be compared to reflexions in rough water on a small scale. The vertical lines of the tree stems, though much blurred, are preserved, and there is no visible reflexion of the horizontal branches of the tree on the left. Much the same sort of effect is given by rain falling on still water.

Owing to the elongation of the image in rippled water it sometimes happens that with the approach of a breeze an object becomes visible by reflexion which in the perfectly smooth water did not appear at all. This might be the case if one were standing on the high cliff at E, Fig. 4. As before explained, one cannot from this point see anything of the house in the smooth water, but if the image were prolonged on the rippled surface to twice or thrice its normal length, it would then become visible by reflexion from the farther sides of the waves. Another curious effect may be noted. In the first chapter attention was

XVII.

Bellagio; a gentle ripple.

XIX.

Killary Bay; more movement of the water.

F. Newington.]

XVIII.

[*Lewes.*

XX.

drawn to the difference in apparent outline between the "reflexion" (page 10) and object reflected. For instance, cows grazing on the side of a hill, across the still water, that are scarcely visible in the twilight, may in the reflexion appear to stand out sharply against the sky-line. If the water becomes ruffled, the reflexions of these cows are drawn out into long streaks. Thus, an inconspicuous object in such a position may often cause in gently rippled water a quite unexpected streak, puzzling the beholder to account for its existence.

A few words may be added on the subject of the painting of reflexions in rippled water. We have seen that, at some distance from the eye, where the elongated images are formed by reflexion from numerous wave-surfaces, they seem to be made up of vertical streaks, whilst nearer at hand these vertical streaks resolve themselves (as explained on page 34, and instanced on page 37 in the case of the moon) into a succession of horizontal lines. It would therefore appear that in those parts of the water too far off for the ripples to be separately visible, the effect can best be imitated by *vertical* strokes of the brush, but in the foreground, where the ripples assume distinct proportions, it must be given by a *horizontal* touch. In Plate VI, page 32, where the ripples are very gentle and regular, the reason for this distinction is evident. The reflexion of the buildings on the opposite shore is dragged out so that it looks like a series of vertical lines, whilst that of the mountain peaks in the foreground is broken up by spaces of sky reflexion on the near sides of the waves, leaving dark horizontal strokes on their further sides.[1] The characteristic of moving water is indeed to insist upon the vertical and ignore the horizontal lines of the image, though very regular ripples in the foreground often seem to contradict this principle by showing horizontal lines that do not exist in the object at all. The reflexion of the sunset sky in Plate XXI affords a striking instance of an apparent exception of this kind, the cloud forms being broken up horizontally by the extremely gentle motion of the water. Plate XXII, in which the ripples are fairly regular and travelling obliquely, shows the characteristic " markings " formed on their farther sides by the reflexion of the dark wooded bank.

But it will, no doubt, have been noticed that these horizontal strokes often take the form of elongated *rings*. In Plate XXIV the reflexion of the distant and nearly horizontal bank is seen to consist of a number of very flat rings. When the spectator's position is not too low, so that the whole surface of a wave is in view, as when looking from Q, Fig. 14, there are generally *two* points on each wave at which a suitably placed luminous point, such as P, can be seen reflected. Thus, on the farther side of the first wave, P appears by reflexion at I and H, and on the second wave, at G and F. Beyond X, the two points are on the near side of each wave, viz., D and C on the fourth, and B and A on the fifth wave.[2] Now if, in

[1] Compare also the rippled surfaces in Plates XII and XIII, page 44, and in Plates XVII and XVIII, page 50.

[2] In a sine curve (page 31) the two points of each pair would be proportionally farther apart than in the circular curve of the diagram.

H. P. Robinson.] [Redhill.

XXI.

SUNSET: NORFOLK BROADS.

Fig. 15, which represents the view of P and its reflexions as seen from Q, we draw through P a horizontal line,[1] we shall get as the reflexions of this line a succession of pairs of horizontal lines (one pair for each wave) through the points *a* and *b*, *c* and *d*, *f* and *g*, *h* and *i*.[2] But, owing to the irregularity of the surface of the water, and the breaking up of the troughs by crossing waves into comparatively short depressions, these different reflexions of a horizontal line cannot continue in the water as horizontal lines. If we follow one of the pairs of theoretically horizontal lines on the side of a wave, we soon come to a place where the trough assumes such a shape that only one image, or none, can be formed; so that the two lines merge into one or vanish completely, perhaps to reappear a little farther on, and instead of two parallel lines we see a chain of loops or a series of disconnected rings.[3]

Such rings are amongst the commonest features of gently moving water in the foreground of a picture. The reflexions of a boom or bowsprit, or of any conspicuous horizontal line, often assume this form (see

On the third wave there is only one point at which P can be seen reflected, viz., E.

[1] Representing a straight line through P, Fig. 14, at right angles to the plane of the paper.

[2] And a single horizontal line through *e*.

[3] A cup filled with water is all the apparatus necessary for the production of "rings." Let the cup be placed so that a horizontal bar of the window frame is seen reflected in the still water. If the water be gently stirred round, its surface will become concave, and when the rotation has nearly ceased, the reflexion of the window-bar in the hollow surface will be seen to take the form of a ring.

Plate XXXV, page 74). In Plate XXIV, at that part of the surface where the inside of the ring appears, sky is seen by a double reflexion. We have already had an instance of this in Fig. 13 (page 32). When looking towards *b*, the line of vision cannot continue in the direction *bg*, but, striking the near side of the next wave, is again reflected, so that sky is seen in the direction *gh*. Fig. 18 shows the extent of the reflexion of the bank PS on the farther side of the first

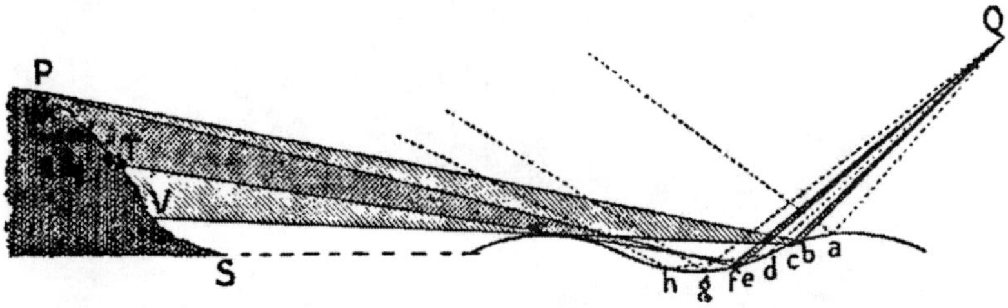

Fig. 18. The point P is seen reflected at two points on the surface of the first wave, viz., *b* and *f*. Nearer than *b*, as, for instance, at *a*, we see sky, as indicated by the dotted line. Beyond *b* we see bank (PV) reflected as far as *c*, where the top of the second wave comes in the way, hiding all the bank below V. Beyond *c*, as at *d*, the line of vision, being further depressed, strikes the side of the next wave at *h*, and is therefrom reflected up towards the sky. *dh* represents about the lowest inclination of the line of vision after reflexion, and beyond *d* it begins to rise again, until at *e* the bank (at T) is again visible over the top of the next wave. From *e* we continue to see bank as far as *f*, where P is again reflected; and beyond *f*, as at *g*, again only sky. In order to give greater clearness to the diagram, the space between the lines Q*b*P and Q*c*V is tinted, and also that between Q*e*T and Q*f*P.

wave when looking from Q. The reflexion of the bank appears only on the two short portions (*bc* and *ef*) of the surface which are darkened in the figure, and between these sky is seen by a further reflexion on the

XXII.

CLIFDEN.

Regular ripples travelling obliquely. A suggestion of rings.

XXIII.

NAUPLIA.

Gentle but complex movement of water. Ghost-like reflexions of white buildings.

near side of the next wave. These two portions of bank-reflexion, if followed in either direction (*i.e.* at right angles to the plane of the paper), will, as explained above, owing to local irregularities, unite sooner or later to form a flattened ring as in the photograph (Plate XXIV). If there were a light space of sky above the bank and above that again a stratum of darker cloud, the dark rings caused by the reflexion of the dark bank would in all likelihood be surrounded by light rings reflecting the light part of the sky above it. In Plate XXXV the rings over the oar to the right are no doubt caused by the boom, their dark centres being probably due to a second reflexion showing the sail.

In near ripples, which, from their proximity to the eye, appear large and few in proportion to the size of the object reflected, the image becomes strangely distorted, dancing madly with the movement of the water. This is particularly noticeable in the case of the foreground reflexions of masts and rigging, which writhe and twist themselves into bends and knots, or even detached rings. (See Plates XXV and XXXII, and note the curious zigzag reflexion of the bowsprit in Plate III, page 25.) In the middle distance such objects are reflected with less distortion, though not so sharply as these in the foreground. The reflexion of the hull of a boat, or even the lower parts of its rigging, takes place on a part of the water much farther off from the spectator than that of the top of the sails, so that, speaking roughly, the distortion of the image of a boat may be said to increase from the water-line upwards (see Plate XXVIII). The excellent photographs of fishing smacks by Mr. C. E. Wanless show some beautiful and curious effects of foreground reflexions, and are of more value than many words in illustrating the play of the water, or as suggestions for the drawing of ripples. (Plates XXVI—XXXII, XXXVII.)

NOTE ON THE PERSPECTIVE OF ELONGATED REFLEXIONS.

On page 36 we saw how the streak of light which appears in rippled water as the reflexion of a luminous point, such as a lamp, is formed by the light catching the surface of the water at different points which, neglecting the irregularities of the surface, may be said to lie on a straight line on the surface extending from a point vertically below the eye to a point vertically below the lamp. For example, the points A, B, C, D, E, F, etc. (Figs. 14 and 15) lie on such a line. If several lamps are seen reflected side by side, the streaks of light on the surface of the water corresponding to them will evidently *not* be parallel, radiating as they do from the point below the eye. But these streaks of light regarded as lines *below* the surface have been shown (page 36) to appear as vertical—and therefore parallel—lines.

The question might therefore arise, should these streaks of light *be drawn* as vertical lines, or as lines radiating from the "Point of Station?" That the former is the correct answer is clear from a consideration of Fig. 14. Viewed from Q, the perspective representations of the points *a*, *b*, *c*, *d*, *e*, etc. (*i.e.*, their projections on to the picture plane), must lie in the same vertical line with that of P, as shown in Fig. 15. If there were a second luminous point alongside of P (Fig. 15) its images would in the same way arrange themselves in the vertical line through that point, and so on for any number of luminous points. Should further proof be required, it is given by the law of perspective that the vanishing point for any given receding line is the point where the line from the eye parallel to it meets the picture plane. It will be found that the projections on to the picture plane of lines in the ground plane radiating from a point vertically below the eye are vertical lines. This is illustrated in Fig. 19. The spectator is here supposed to be stationed above the water and to be looking directly at the bridge to the right of the diagram. E marks the position of his eye or "station point," A B C D represents his "picture plane," with the projection of the bridge as it appears to him, *o* being the "centre of vision," E*o* the "direction of vision," and H the horizon line. EX is a perpendicular drawn from E to meet the "ground plane" (surface of the water) in X. L_1, L_2, etc. are lamps on the near side of the bridge, and F_1, F_2, etc., are points on the surface of the water vertically beneath these lamps. The streaks of light on the surface of the water caused by these lamps will appear to one looking from E to lie along the straight lines XF_1, XF_2, etc., which radiate from the point X immediately beneath his eye. But the projection of these radiating lines on to his picture plane will be vertical, and therefore parallel, lines. Let the lines XF_1, XF_2, etc., cut the base of the picture plane (or "ground line") in the points g_1, g_2, etc. From these points draw perpendiculars $g_1 h_1$, $g_2 h_2$, etc., to meet the horizon line of the picture in h_1, h_2, etc. Then the lines EX and h_1g_1, being vertical, are parallel to each other; and since the plane containing E and the horizon is parallel to the ground plane, these lines, EX and h_1g_1, are also equal. Therefore Eh_1, is parallel to Xg_1; and therefore, by the well-known law of perspective, h_1 is the vanishing point for the line XF_1. Similarly it can be shown that h_2 is the vanishing point for XF_2. h_3 for XF_3, and so on. Thus in the picture the streaks of light are represented by the vertical lines $f_1 g_1$, $f_2 g_2$, $f_3 g_3$, etc.

In "Modern Painters"[1] Ruskin makes reference to a picture by Claude, in which the artist has fallen into an error

[1] Vol. I, Part II, Sec. V, Chap. I, § 17.

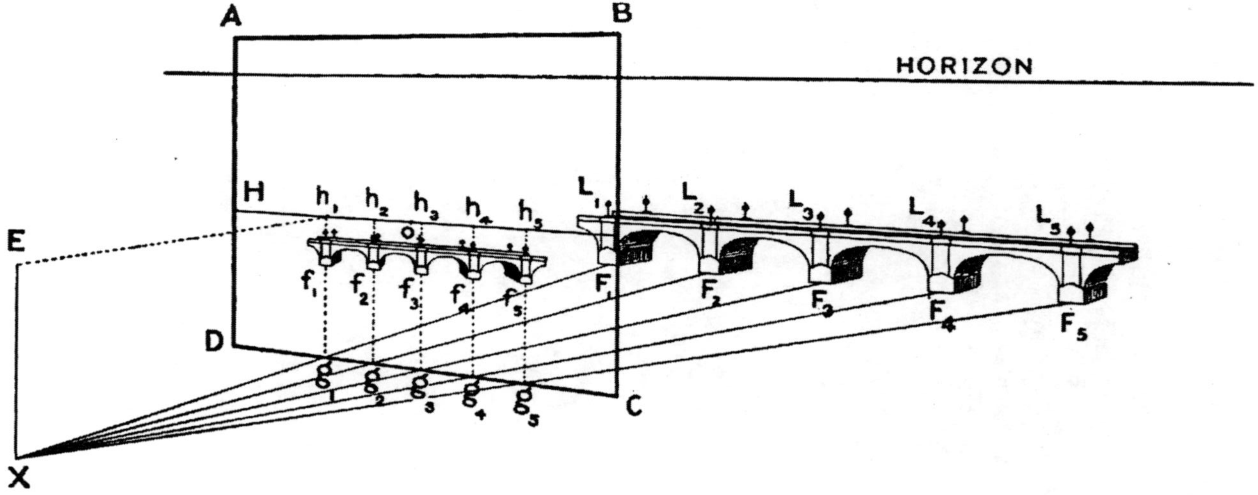

Fig. 19. Looking from E the reflexions of the lamps L_1, L_2, etc., in the rippled water appear as *radiating* lines of light XF_1, XF_2, etc. The projections of these radiating lines on to the picture plane (A B C D) are the *vertical* lines $g_1 f_1$, $g_2 f_2$, etc. The streaks of light representing the reflexions of lamps must therefore be drawn as parallel, and not as divergent, lines. The like error of drawing the path of light beneath the sun narrowing in perspective in the distance (as if it were a real path on the surface of the water) is also to be avoided; see Plate IX, page 38.

XXIV.

LOCHINVER BAY.

The reflexion of the bank opposite takes the form of flattened rings. (In order to show these rings clearly, it was found necessary to darken them by hand, with the result that they appear unduly black ; but nothing has been added that could not be seen in the original photograph. The light ring in the immediate foreground due to sky

through ignorance of this principle. "In one of the smaller rooms of the Uffizi at Florence, off the Tribune, there are two so-called Claudes; one a pretty wooded landscape, I think a copy, the other a marine with architecture, very sweet and genuine. The sun is setting at the side of the picture, it casts a long stream of light upon the water. This stream of light is oblique, and comes from the horizon, where it is under the sun, to a point near the centre of the picture. If this had been done as a license, it would be an instance of most absurd and unjustifiable license, as the fault is detected by the eye in a moment, and there is no occasion nor excuse for it. But I imagine it to be an instance rather of the harm of imperfect science. Taking his impression instinctively from nature, Claude usually did what is right and put his reflection vertically under the sun; probably, however, he had read in some treatise on optics that every point in this reflection was in a vertical plane between the sun and spectator; or he might have noticed, walking on the shore, that the reflection came straight from the sun to his feet, and intending to indicate the position of the spectator, drew in his next picture the reflection sloping to this supposed point, the error being excusable enough, and plausible enough to have been lately revived and systematized." The gist of the whole matter is contained in a footnote. "Every picture is the representation of a vertical plate of glass, with what might be seen through it drawn on its surface. Let a vertical plate of glass be taken, and wherever it be placed, whether the sun be at its side or at its centre, the reflection will always be found in a vertical line under the sun, parallel with the side of the glass. The pane of any window looking to sea is all the apparatus necessary for this experiment; and yet it is not long since this very principle was disputed with me by a man of much taste and information, who supposed Turner to be wrong in drawing the reflection straight down at the side of his picture, as in his 'Lancaster Sands,' and innumerable other instances."

CHAPTER III

COLOURS IN STILL WATER

WE have considered so far the reflexion of light at the surface of water. But not all the light that strikes the surface is reflected, for some of it passes on into the water; so that, in order to understand how it is that we see the real colour of water,

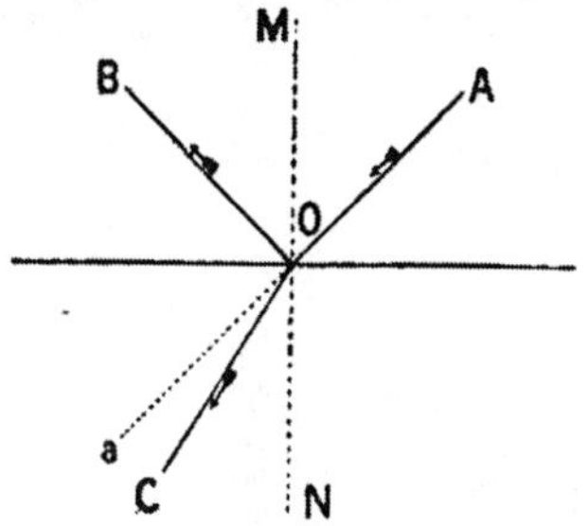

Fig. 20. Refraction of a ray of light on entering water.

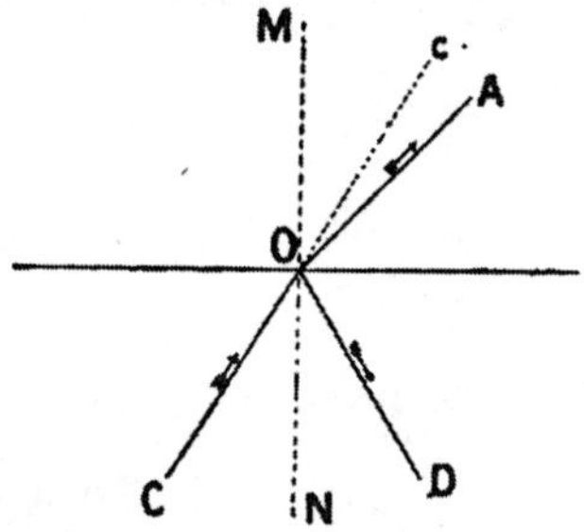

Fig. 21. Refraction of a ray of light on emerging from water.

as distinguished from that due to surface reflexion, we must trace the passage of light from air into water, and *vice versâ*, from water into air.

Suppose a beam of light AO (Fig. 20) to strike the surface of still water at O. A portion of the light will then be reflected "regularly" (as explained in Chapter I) along OB and the remainder will enter the water. This second portion will not continue its course in a straight line to *a*, but will be

F. M. Sutcliffe.] [*Whitby.*

XXV.

Knotted reflexions of masts.

C. E. Wanless.] [*Scarborough.*

XXVI.

Broken reflexion of sail. The mast, being taller, is reflected as a continuous winding line.

"refracted," or bent downwards, in the direction OC. Fig. 21 is similar to Fig. 20, and shows the passage of light in the opposite direction. In this case the ray is supposed to start from below the surface of the water at C; on reaching the surface at O, part of it is reflected downwards along OD, and the rest emerges, not in a straight line to *c*, but bent in the direction OA.[1] The ray thus follows the same path as that starting from A in Fig. 20, but in the opposite direction.

This bending or refraction of light in passing from

[1] The angle of refraction, CON (Fig. 20), is measured, like the angles of incidence and reflexion, from the normal at O, MN. The laws of refraction of light are as follows:

(1) *The refracted ray lies in the plane containing the incident ray and the normal, and on the opposite side of the normal.*

(2) *The sines of the angles of incidence and refraction always bear a constant ratio to one another, called the index of refraction.*

For light passing from air into water, $\frac{\text{sin angle of incidence}}{\text{sin angle of refraction}} = \frac{4}{3}$, or in other words, the *refractive index of air to water* is $\frac{4}{3}$ or 1·33.

If the ray CO is more inclined than in Fig. 21, it is evident that it will reach a position at which the refracted ray OA will lie along the surface of the water. In this case, the angle MOA becoming 90°, sin MOA = 1; therefore sin CON = $\frac{3}{4}$ and CON = about $48\frac{1}{2}$°. This is known as the *critical angle* (KON, Fig. 22); and if the incident ray is still further inclined, as, for instance, to the position CO in Fig. 22, there will be no refracted ray, but the whole of the light will be reflected downwards in the direction OD. This is called *total internal reflexion.*

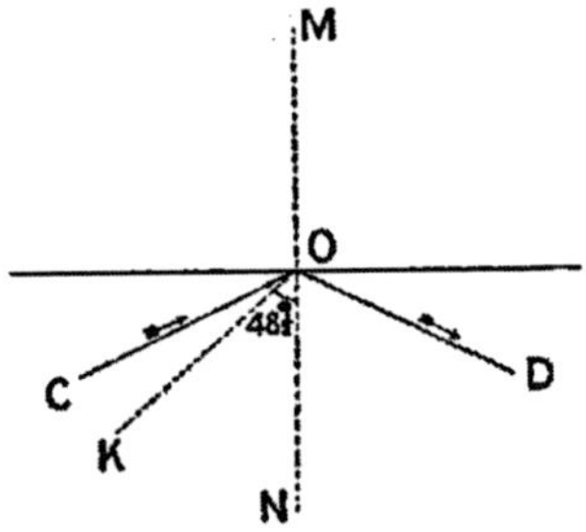

Fig. 22. Total internal reflexion.

air into water, or *vice versâ,* is a matter of everyday experience. It is a well-known fact that a coin just out of sight at the bottom of a cup can be made visible by filling the cup with water. A ray of light from a point C on the coin (Fig. 23) cannot travel in a straight line from C to the eye at A, but must follow a bent path, as in Fig. 21. The ray CO is bent downwards on leaving the water, and thus enters the eye; and the eye, assuming that light travels in straight lines, sees the coin in the direction AO at *c*, so that the water appears to be shallower than it actually is.[1] In the same way a stick thrust obliquely into water seems to be broken or sharply bent upwards at the point

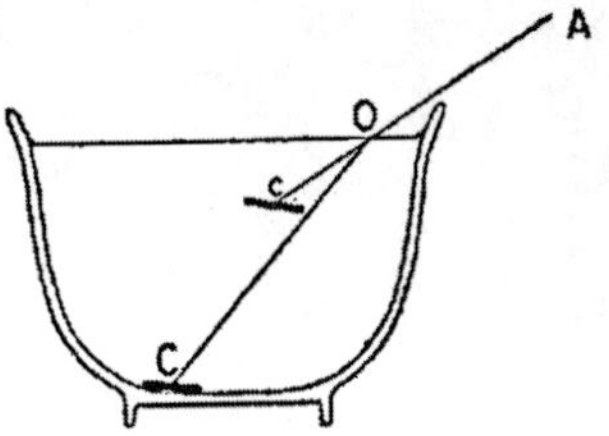

Fig. 23.

[1] The image of a point formed by refraction is not stationary, as in the case of one formed by reflexion, but varies with the position of the eye. In Fig. 24 (taken from Prof. W. Watson's "Text Book of Physics," Longmans, Green and Co., 1899) the points *p*, *p′*, *p″*, which lie on a "*caustic curve,*" mark the different positions of the image of the point P, corresponding to the different positions of the eye, *a*, *a′* and *a″*. So that the more oblique the direction of vision, the shallower the water appears to be. A sunken rock, that can easily be passed over in a boat, appears, as we approach it, to reach within a few inches of the surface of the water; but looking vertically down, the water seems some three-fourths of its actual depth.

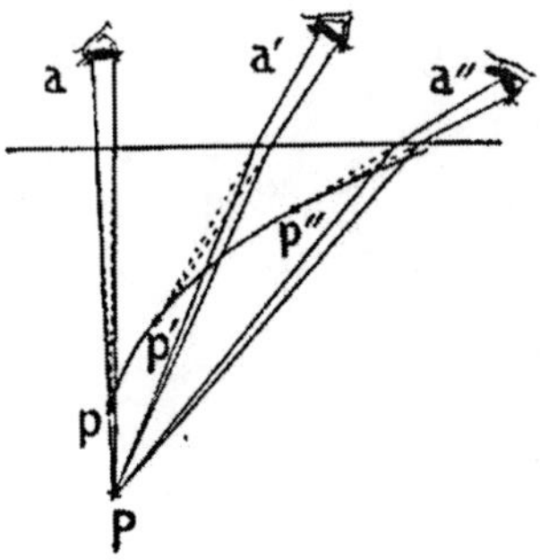

Fig. 24.

C. E. Wanless.] [Scarborough.

XXVII.

OLD HARBOUR SIDE, SCARBOROUGH.

where it enters the water. The angle of refraction, as explained in the note above, bears a certain definite relation to the angle of incidence. That which at present concerns us is, however, not the relation of these angles, but the amount of light that passes from air into water compared with the amount that is reflected at the surface; and again, the amount that passes out of water from a point beneath the surface compared with the amount that is reflected down

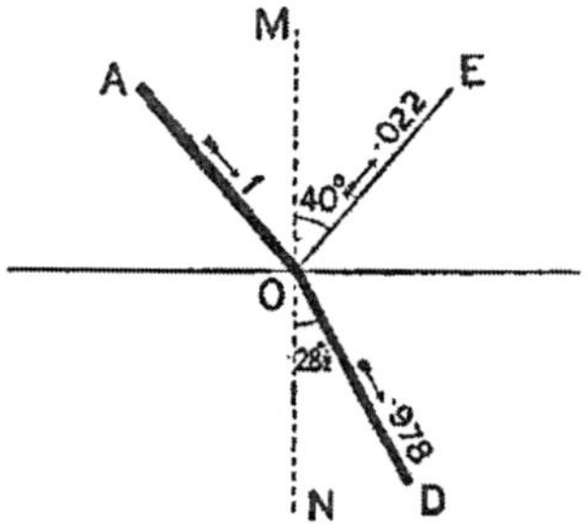

Fig. 25 (I).

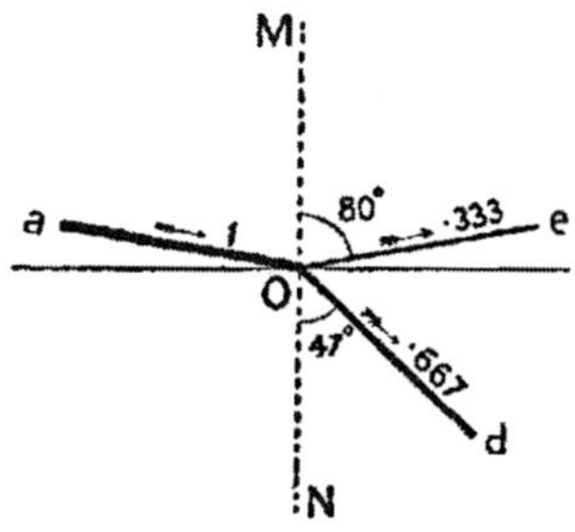

Fig. 25 (II).

again into the water. We shall find that the more obliquely the ray falls on the surface, the more of it will be reflected and therefore the less refracted, and *vice versâ*, the more perpendicularly it falls on the surface, the more will enter the water and the less be reflected.

This principle is roughly illustrated in Fig. 25 by the relative thicknesses of the lines representing the rays of light. For example, in Fig. 25 (I) almost the whole of the ray AO (inclined at an angle of 40°) is refracted along OD, only a very small fraction of the light (about one-fiftieth) being reflected, whereas in (II), where the ray *a*O strikes the surface very obliquely (at an angle of 80°), a much larger portion (about one-third) escapes at the surface and is reflected along O*e*.[1] Figs. 25 (III) and (IV) are similar to (I) and (II) respectively, the light now travelling in the direction of the arrows from a source below the surface, and here in the same way a much larger proportion of the beam CO passes out of the water than of the more oblique beam *c*O, which is nearly all reflected down again to *d*.

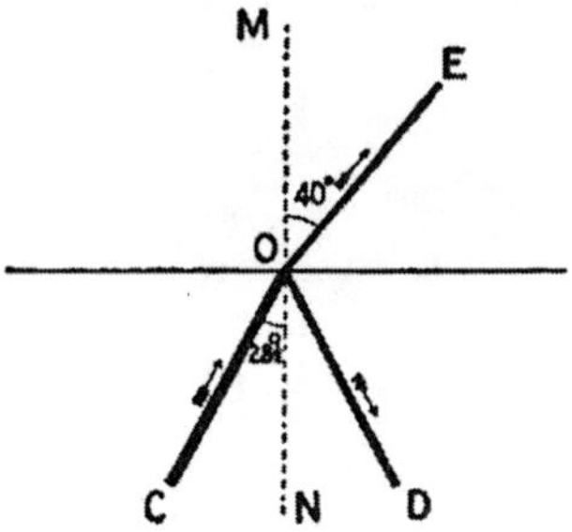

Fig. 25 (III).

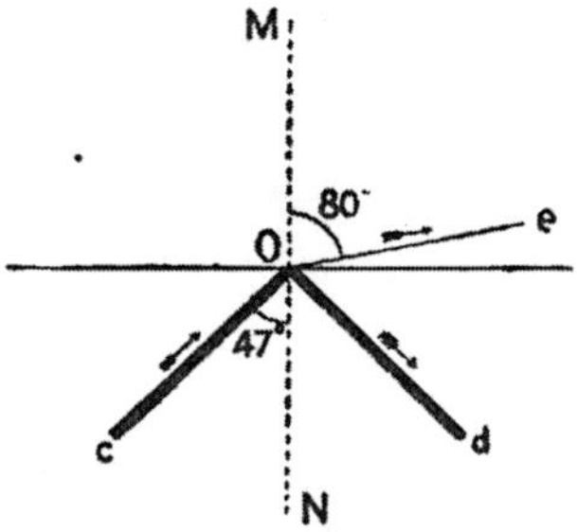

Fig. 25 (IV).

If Figs. 25 (I) and (III) were placed one on the top of the other, the point O on the point O, and the line MN on the line MN, then the lines OE, OD, in each diagram would also coincide. A combination of the two diagrams would therefore represent the light coming to the eye, if placed at E and directed

[1] The following figures are taken from "Light," by Professor Tyndall. At perpendicular incidence,

(*i.e.*, at 0°)	water reflects		18	rays	per	1,000
at 40°	"	"	22	"	"	"
at 60°	"	"	65	"	"	"
at 80°	"	"	333	"	"	"
whilst at 89½°	"	"	721	"	"	"

C. E. Wanless.] XXVIII. [*Scarborough.*

C. E. Wanless.] XXIX. [*Scarborough.*

A gentle swell. (Where sky alone is reflected, the near sides of the waves may be distinguished from the farther sides by their darker shade.)

towards O. The ray OE would then consist of rays travelling along AO and CO, and—assuming the water to be still—of these rays alone. It is evident that only a minute fraction of the light from above reaches the eye, whilst it receives a very considerable proportion of the light from below the surface. On the other hand, by superimposing Figs. 25 (II) and (IV), we see that when the eye is placed at *e*, so that the line of vision, *e*O, strikes the surface of the water very obliquely, it receives a large amount of reflected, and only a small portion of refracted, light. The significance of this point will soon be apparent.

When light is reflected at the surface of water it is unchanged in colour, and therefore the "double" of the objects that we imagined inverted below the surface of the water in Chapter I, is similar to the real scene in colour, as well as in form, though, as our view of this image is a different one to our direct view of the objects, so we may get in the reflexion different combinations of colouring or different effects of light and shade (page 17). But besides the reflexions that we see on the water, we actually perceive in most cases and to a certain extent the colour of the water itself. Let us consider this apart from the reflexions. How do we get the sensation of colour? We know that by passing the sun's light through a prism it can be split up into rays of different wave-lengths, which produce upon the eye the effects of the different colours of the rainbow, and that these coloured rays can be reunited by means of a second prism to form white light again. So we say that white light is composed of light of all colours. Any transparent coloured substance, placed so as to intercept a beam of white light, only allows light of certain wave-lengths to pass, which give the sensation of colour to the eye. Thus in looking through a piece of red glass we see red, because all (or nearly all) the rays except the red ones are absorbed.[1] By experiment with the prism it will at once be seen that the red glass does not convert white light into red, but that it produces the change by subtraction of some of its constituents. Again, in the case of most opaque coloured objects, part of the white light falling on the object is reflected at the surface unchanged, while the remainder penetrates to a small depth below the surface before being reflected, and on its passage through a thin layer of the coloured substance suffers absorption of some of its component parts and emerges as coloured light.

It is evident then that in the case of the red glass we shall see no colour unless light pass through the glass. Place a piece of coloured glass on a dull black cloth and it appears black; but lay it on a sheet of white paper and its colour at once becomes visible. The paper reflects light from its surface through the glass, which the black cloth does not. So it is with water; if a white stone is dropped into a clear lake, the deeper it goes the more it shows the real colour of the water—generally a greenish-blue—as the light reflected from it has to travel farther through the

[1] This is true of the red glass used by photographers to exclude actinic rays; but it is not necessary that all light of other colours should be absorbed by the medium to give the sensation of red. If the complementary colour alone is absorbed we see red, though this red is actually far from being pure.

C. E. Wanless.] XXX. [Scarborough.

C. E. Wanless.] XXXI. [Scarborough.

water. If the water is deep enough, the stone disappears, for the blue rays, though not absorbed by the water so quickly as the others, cannot pierce the great distance from the surface to the bottom and back again. So that if the water is clear, that is to say, if it carries no particles to reflect light to the eye, and is sufficiently deep, it will appear black.[1]

But in nature water is seldom quite clear, and light is usually reflected upwards from suspended particles, and if shallow, from stones at the bottom also. Thus it is that we see the colour of water. Light from above enters the water and is reflected to the eye from the minute particles floating in it; in passing through the water, certain of its constituent rays are absorbed in greater proportion than others, so that it has on exit a predominance of blue or green rays. The resultant colour depends of course to some extent on the colour of the particles themselves. White particles floating in pure water[2] make it appear blue, whilst yellow particles give it a greenish tint.

In lakes the blackness is probably, in most cases, due to yellow stain, which stops all the light escaping the absorption of pure water.

[2] Without going here into the question of the inherent colour of water, it may be stated generally that *pure* water is of a beautiful blue colour, very closely resembling that of Prussian blue. When stained with peat, or other vegetable matter, it becomes first green, then yellowish-brown. Thus the difference in colour, often so marked, of the Swiss lakes is accounted for, the bluest lakes being the purest, whilst the greener lakes contain a small quantity of vegetable matter in solution. Again, with regard to the seas surrounding our islands, though their greenness may be in part due to the yellow colour of the minute particles of floating sand, it is probably also largely caused by the influx of yellowish-brown river

In any sheet of coloured water, therefore, the apparent brilliancy of its colour depends upon *the amount of light coming to the eye from beneath the surface*, the light which is reflected at the surface being, as we have said, unchanged in colour.

Now this amount obviously depends first on the amount of light which *enters* the water. For example, the sea on our coasts is generally of a greenish hue. This colour often reveals itself in the thin edges of the waves where they rise steeply or curl over before breaking, but, except where the crests become thus translucent, it is not always evident. With a cloudy sky comparatively little light enters the water, and therefore comparatively little is reflected from the floating particles of sand. The amount of greenish light reaching the eye is thus very small compared with the amount of gray light from the clouds seen by surface-reflexion, and the water looks a dull gray. If, however, the clouds part and a beam of direct sunlight strikes the water, the eye is relieved by a gleam of emerald.

Secondly, it depends on the proportion of suspended particles. As we have seen, if the water is perfectly clear and very deep, it will look black, all light that enters it being absorbed. But where suspended particles are present, as explained above, the colour is brought out, and, up to a certain point, the more particles the more colour. Beyond this point the water commences to become turbid, and gradually loses its characteristic green colour, as the colour

water. (Dr. John Aitken, in "Nature," Vol. LIX, page 509). See Note at end of chapter, page 82.

C. E. Wanless.] [Scarborough.

XXXII.

SCOTCH FISHING BOAT, SCARBOROUGH.

of the suspended particles in increasing number overpowers that due to the water itself. It is hardly necessary to add that the colour of thick or muddy water is not the colour of the water at all, but of the solid particles it carries. Thus the Upper Engadine lakes, the water of which is a fine blue-green,[1] owe the brilliancy of their colouring to the unceasing supply of minute white particles brought down by the glacier-fed torrents. These lakes, which are four in number and connected by short lengths of river, present a distinct gradation in colouring from the upper to the lower. For it is at the head of the valley that there are the greatest number of affluent streams; and the uppermost lake, the Silser See, receives therefore more than its proportion of glacier dust. The water in passing through this lake deposits a great deal of the dust as sediment, only the finest particles being carried on to the next lake, so that the proportion of suspended particles diminishes from lake to lake. Hence the corresponding gradation in brilliancy of colouring. In early summer, when the glacier streams are swollen and turbid, the upper lakes become almost milky in appearance; but by the time the river has reached the St. Moritz lake, the last of the series, it has dropped most of its glacier dust, retaining just enough to display the full beauty of its blue-green colour, though it is perhaps surpassed in brilliancy by the little Campfer lake above. In the sea similar changes of colour accompany changes in

[1] The colour of the water of these lakes is seen to advantage in the blocks of ice sawn out of them in winter, which will doubtless be found somewhat greener than the pure blue of a glacier crevasse.

the relative amount of chalk powder or fine grains of sand floating in the water. Near the shore or over a sand bank it is almost always greener than in deep water, though close to the beach it may often, owing to excess of floating particles, assume a milky or a sandy colour. The colour of sea-water is also well brought out by foam, or sunken particles of air, as in the wake of a screw steamer.[1]

To repeat: the brilliancy of the colour is dependent, first, on the illumination from above, in other words, on the condition of the sky: secondly, on the condition of the water with regard to suspended particles. Having got these two very obvious conditions, let us inquire how it depends on the position of the observer. As in the consideration of reflexions in Chapter I, we will first suppose the water to be absolutely smooth.

Leaving for a moment the question of colour, let the reader imagine himself to be standing on a stone lying in the shallow water at the edge of a lake. The water is perfectly still and clear, and masses of white cloud are floating overhead. Looking down at his feet, he sees two pictures; (1) the bottom of brown mud or yellow gravel, and (2)—though he may not at first have noticed them—the sky and clouds also. These two pictures, the one caused by rays coming from beneath the surface, and the other by rays reflected at the surface of the water, are super-

[1] In very shallow water the brilliancy of colour depends of course on the reflective power of the bottom. The "silver sand" that occurs on some parts of our coasts, possesses this power in a high degree.

imposed one upon the other, though, by reason of the great difference in distance of the objects forming them, it is not easy to look at both at once. The clouds, being so far off, demand a different adjustment of the eyes to that needed for the objects at his feet, so that, on looking from one picture to the other, he is conscious of an actual movement of the eyes. Now if, instead of looking straight down into the water, the line of vision be gradually inclined, the one picture will be found to gain in strength, and the other to fade rapidly away; in other words, the reflexion of the sky becomes more distinct, whilst the bottom disappears. Thus the principle illustrated in Figs. 25, pp. 63 and 64, is verified, viz., that the more obliquely one looks at the still surface of a sheet of water, the more will the eye receive of reflected and the less of refracted rays. As the gaze is turned farther away from the feet, the picture of the sky formed by reflexion, which has the great advantage of being far the more luminous, predominates so rapidly over the other that at a distance of a few yards, say ten, it has entirely supplanted it. It might be objected that the increased depth of water would alone account for this, but the change will be found to be just as remarkable, or very nearly so, on looking in a direction parallel to the shore, where the depth remains practically the same.

Exactly the same principle applies in the case of deeper or less clear water, where, looking down at it, we do not see the bottom, but we do see the colour of the water. We have, as before, the picture of the sky formed by rays reflected at the surface superimposed upon the other picture, formed by rays coming from beneath the surface, which latter, proceeding in this case not from the bottom, but from particles floating in the water, give merely a general impression of its "local colour;" and, as before, the more obliquely we look upon the water, the more the former picture gains in strength at the expense of the latter.

So that we have this further and more important condition governing the apparent colour of water. Given smooth water of a definite colour, with a fixed proportion of suspended particles, and illuminated to a certain degree, *the apparent strength of colour depends upon the position of the observer, being greatest on looking vertically down at the surface and lessening gradually as the line of vision strikes the surface more and more obliquely, the colour due to reflexion gaining in brilliancy as the local colour dies away.* For instance, when looking very obliquely at a sheet of water in the distance, the objects beyond are reflected as in a mirror, the reflexions being practically identical in colour with the objects themselves, whilst the water itself appears perfectly colourless.[1] Indeed it might be turned into blood and we should not perceive the change. The same phenomenon can be observed in any flat reflecting surface, such as the top of a well-polished table. Sitting in a low chair, with the eye only slightly above the level of the table, one sees the objects beyond distinctly reflected, and one could not

[1] On referring to the table in the Note, page 64, it will be seen that at very oblique incidence water reflects by far the greater part of the light that falls upon it. At the same time the amount of *refracted light* reaching the eye in this position is so small as to be negligible.

(without previous knowledge) say whether the surface were of white marble or of polished ebony. Or a piece of coloured glass may be placed on a sheet of white paper, and it will be found that, as the glass is turned sideways, a position is reached in which its colour can no longer be seen. In the same way, wet patches of smooth sand left by the ebbing tide, when viewed sufficiently obliquely, show only sky reflexion, and nothing of the actual colour of the sand.

On the other hand, we get the opposite extreme, if, standing at the edge of a muddy duck-pond in full sunshine, we look directly at the near surface of the water. The local colour, which resembles that of green pea-soup, is now quite conspicuous, especially if the pond has been lately stirred up by its tenants or by a heavy shower of rain, whilst the reflexion-picture at this abrupt angle is very faint, though it rapidly asserts itself as we look more obliquely at the water. The same thing is true in a less degree of the sea on a calm day, when the greenish colour of the nearer water is far less apparent from the beach than from the top of a high cliff.[1]

[1] The following simple experiment will serve as a practical illustration of this principle. Take a plain white basin, tip it up on one side, as in Fig. 26, and fill it as full of water as is possible in this position. Now take a small object, such as a gold stud or black button, attached to a piece of thread. Look vertically down into the water, hold the button by the thread, so that it hangs immediately beneath the eye, and lower it into the water. Before it reaches the surface its reflexion will only be very faintly seen, and after entering the water it will itself lose very little in brilliancy, thus showing that in this position (*i.e.*, when the line of vision is vertical) the refracted ray reaching the eye is much stronger than

These instances show how much depends upon the angle of vision, and the importance of taking it into account in painting reflexions. If we look at the image of the reflected trees, we shall see that those parts of the foliage which appear in the immediate foreground are modified by the colour of the water itself (whether it be the pure blue of the Swiss lakes, or the muddy bottom and slimy water of a duck-pond), that this local colour shows less in the foliage reflected a little farther off and disappears altogether in the distant reflexions. This gradation is so subtle that it may easily pass unobserved, but, if we want to be true to nature in our colouring, we must not omit to give it expression. The difference of colour

the reflected ray. Now, standing on the side towards which the basin is tipped up, lower the eye almost to the level of the water, and again let down the button as before (Fig. 27). The effect in

Fig. 26.

Fig. 27.

this case will be very different; the reflexion of the button before it enters the water will be nearly as distinct as the object itself, while the button after entering the water almost entirely disappears. Thus it is evident that on looking very obliquely on to the surface,

XXXIII.
WHITBY.

XXXIV.
MULBARTON.

The amount of local colour seen in water depends upon the angle of vision.

F. M. Sutcliffe.] [*Whitby.*

XXXV.

"RINGS" (see page 53).

is, however, very obvious in a case like that shown in the accompanying photograph of boats, taken from the high edge of the quay at low tide (Plate XXXIII). In the reflexions of the first row of boats the dirty green of the water is far more apparent than in the reflexions of the second row, so much so that the bright stripes of paint on the nearer boats are almost lost, whilst those on the boats beyond are distinctly reproduced. Plate XXXIV is a further illustration of the same point. In this case the colour of the water could be plainly seen in the reflexion of the nearest duck, whilst that of the farthest showed no sign of it, but looked as white as the duck itself.

It need scarcely be pointed out that, speaking generally, the "value" of water in the foreground,

the reflected ray is much stronger than the refracted. By rough experiment the strengths of the reflected and refracted rays are found to be equal when looking at an angle of about 13° or 14° with the surface.

If we mix into the water some Prussian blue or other pigment, we see its colour by means of light (reflected from the bottom of the basin) which has lost certain rays by absorption during its passage through the blue water, but, on lowering the eye to the second position, the colour of the water entirely disappears, and we see nothing but the reflexion of the white edge of the basin. We may now vary the first experiment as follows:—take a piece of bright red paper and hold it upright with its lower edge almost touching the water on the further side of the basin. The reflexion of this red paper as seen from the second position is almost identical in colour and strength with the direct view of it, but, on slowly raising the head, the blue colour of the water becomes more and more apparent, until, looking abruptly down on to the surface of the water, the red has almost entirely disappeared; so that the reflexion of the red paper seems to change from red through purple to blue. (See also footnotes, pages 79 and 92.)

reflecting as it does at this abrupt angle so much less light from the sky, is considerably below the value of that part of the sky which it reflects. One hears it said that water should be painted "a tone lower" than the sky. As a matter of fact, they are in the distance almost equal in value; whilst, on looking down perpendicularly at the water, the difference in value between them is enormous, the blue sky appearing almost black by reflexion. Thus there is a marked gradation in (diminishing) value from the far distance to the foreground, increased by the fact that the upper parts of the sky are darker than those near the horizon. This gradation may be clearly seen in any photograph which includes a water foreground, as, for instance, in the frontispiece to this volume.[1]

So much for the angle of vision; but the degree of illumination is also an important condition. It has been already noted (page 68) how the colour of water is brought out by direct sunlight, and the difference in the appearance of our duck-pond in sunshine and in shade is very considerable. On a dull day, the colour of the water may perhaps be scarcely perceptible a yard or two from the point at which the observer is standing, and looking more obliquely on to the water, it disappears altogether, so that at the further side of the pond the sky seems to be mirrored in all its purity

[1] Owing to the limited range of luminosity in a photograph, the brightest part of the sky in this view appears no brighter than its reflexion, though the actual difference in tone must have been as great there as in the darker parts of the sky on either side of it. A more natural effect might possibly have been obtained by cutting away the mesh in this block in the brightest part of the sky.

XXXVI.

WAGGONERS' WELLS, HINDHEAD.

Water much discoloured after heavy rain. The reflexions are quite clear in the shadow, but nearly disappear in the sunlight beyond.

of colour. But if the sun is high and shining brightly on the water, the whole pond (except in the shadows) shows its greenish colour, and that chiefly in the dark tree-reflexions. The reflexion of the tree-stem on the farther side of the pond seems, when the sun is hidden, to be identical in colour with the real stem, but as soon as the sunlight strikes the water again, its gray is mingled with green. Wherever a shadow falls across the surface, the colour of the water is less prominent, and therefore the reflexion-picture there is more prominent.

So that in drawing shadows on still water, we must make the reflexions seen in the shadows clearer and purer in colour than those seen outside them. It follows that if the water is very clear and practically colourless, there will be hardly any noticeable difference between shadow and sunshine. Indeed we may safely say that no difference would be visible were it not for minute particles of dust—or even seeds or leaves—that may be floating on the surface. On the other hand, when the water is very turbid, the shadows falling upon it are quite obvious. Plate XXXVI is a photograph of a very muddy pond in bright sunlight. The outline of the shadow of the tree in the foreground is clearly defined. Within this shadow the image of the trees on the opposite bank appears nearly as plainly as it would in clear water; but beyond the shadow, that is to say, wherever direct sunlight falls on the pond, the muddy colour of the water is so strongly brought out that the reflexions, or at any rate the darker parts of them, are scarcely distinguishable. In the English Channel, when, as is often the case after stormy weather, the sea assumes an almost milky appearance owing to the great quantity of minute particles of chalk floating in the water, the shadows cast by passing clouds are very conspicuous.

As the local colour is revealed by strong illumination from above, so it loses in proportion as the reflexions with which it has to contend, are themselves brilliant. Everyone is familiar with the difficulty of seeing objects in the dusk through the closed window of a well-lighted room, until a blind or curtain has been drawn to cut off the bright reflexions on the window-pane; and, in the same way, strong reflexions on the surface of water prevent us from seeing beneath the surface. We have observed that, if the sun is hidden, the colour of the water near us is often barely perceptible; and this is in great measure due to the fact that the reflexion-picture here consists (in all probability) of blue sky or highly luminous clouds; and we also found that the colour of the water was most apparent in the reflexions of the trees, for the reflexion-picture there being dark or feebly illuminated, the colour of the water had again a chance to show itself. Hence the explanation of the well-known fact that it is in the reflexions of dark objects, such as heavy foliage or the tarred side of a boat, that we most clearly perceive the colour of water.[1] This fact is often expressed by saying that bright objects are reflected in their natural colours, whilst those that are very dark appear the colour of the water—a say-

[1] Or, if the water be very shallow, that the bottom is most plainly visible.

ing that must not be taken too literally. For the reflexion in the foreground of even a white object, as, for instance, that of the nearest duck in Plate XXXIV, may be strongly tinged with the colour of the water, and, on the other hand, when looking very obliquely, this colour is quite invisible, even in the reflexions of very dark objects. But within certain limits of the angle of vision it is true enough. Though at the other side of the water we see no green in the reflexion of the sky, we do see it distinctly in the reflexion of the dark wall beyond. The image of a black sail shows much local colour, that of a white sail alongside of it hardly any. The effect can also be imitated, as on page 73, by means of a piece of coloured glass. At an oblique angle, when the red or blue colour of the glass cannot be seen in the reflexion of the sky, it is still visible in the reflexions of dark objects, as the bars of the window frame. No one who has ever been rowed into a sea-cave can have failed to notice as he enters the unusual brilliancy of the colour of the water within it. In this case practically all light is cut off from above, so that the rocky bottom, illuminated from the opening behind, shows off the colour of the water to the greatest advantage.[1]

If we wish to find an object lying at the bottom of

[1] As a continuation of the experiment on page 75, we may look (from the second position, Fig. 27), at the reflexions of different coloured articles, such as books of various colours and shades, and we shall still find that the darker the object, the more the colour of the water shows in the reflexion. The image of a dark brown or dull green book appears much changed by the admixture of the blue of the water, whilst a light-coloured object is repeated almost exactly in the reflexion.

a stream, we require three conditions to help us: 1st, to look down into the water as vertically as possible; 2nd, bright sunshine to light up the bed of the stream; and 3rd, to be able to get into such a position that the reflexion of the bright sky is cut off by the image of a tree or other dark object. So we see that the two pictures presented to us when we look at the water, the one formed by light reflected at the surface, and the other by light coming from beneath the surface (whether the latter consist of stones, weeds or other objects distinctly visible beneath the surface, or only a general impression of colour), are, as it were, always contending with each other for pre-eminence, and that each is largely dependent, not only on the angle of vision, but also on the amount of its illumination as compared with that of its rival.

The whole ground covered by this chapter is so admirably expressed by Ruskin in his "Elements of Drawing,"[1] that we cannot do better than quote his words as far as they serve to illustrate the points we are here concerned with.

"When you are drawing shallow or muddy water, you will see shadows on the bottom, or on the surface, continually modifying the reflections; and in a clear mountain stream, the most wonderful complications of effect resulting from the shadows and reflections of the stones in it, mingling with the aspect of the stones themselves seen through the water. Do not be frightened at the complexity; but, on the other hand, do not hope to render it hastily. Look at it well, making out everything that you see, and distinguishing each

[1] Sixth edition, pages 179-182.

component part of the effect. There will be, first, the stones seen through the water, distorted always by refraction, so that, if the general structure of the stone shows straight parallel lines above the water, you may be sure they will be bent where they enter it; then the reflection of the part of the stone above the water crosses and interferes with the part that is seen through it, so that you can hardly tell which is which; and wherever the reflection is darkest, you will see through the water best,[1] and *vice versâ*. Then the real shadow of the stone crosses both these images, and where that shadow falls, it makes the water more reflective, and where the sunshine falls, you will see more of the surface of the water, and of any dust or motes that may be floating on it: but whether you are to see, at the same spot, most of the bottom of the water, or of the reflection of the objects above, depends on the position of the eye. The more you look down into the water, the better you see objects through it; the more you look along it, the eye being low, the more you see the reflection of objects above it. Hence the colour of a given space of surface in a stream will entirely change while you stand still in the same spot, merely as you stoop or raise

[1] "For this reason it often happens that if the water be shallow, and you are looking steeply down into it, the reflection of objects on the bank will consist simply of pieces of the bottom seen clearly through the water, and relieved by flashes of light, which are the reflection of the sky. Thus you may have to draw the reflected dark shape of a bush, but, inside of that shape, you must not draw the leaves of the bush, but the stones under the water; and, outside of this dark reflection, the blue and white of the sky with no stones visible" (*Ibid.*, page 332).

your head; and thus the colours with which water is painted are an indication of the position of the spectator, and connected inseparably with the perspective of the shores. The most beautiful of all results that I know in mountain streams is when the water is shallow, and the stones at the bottom are rich reddish-orange and black, and the water is seen at an angle which exactly divides the visible colours between those of the stones and that of the sky, and the sky is of clear, full blue. The resulting purple, obtained by the blending of the blue and the orange red, broken by the play of innumerable gradations in the stones, is indescribably lovely. All this seems complicated enough already; but if there be a strong colour in the clear water itself, as of green or blue in the Swiss lakes, all these phenomena are doubly involved; for the darker reflections now become of the colour of the water. The reflection of a black gondola, for instance, at Venice, is never black, but pure dark green. . . ."

NOTE ON THE COLOUR OF WATER.

The *selective absorption* theory of the colour of water is now generally accepted in preference to the *selective reflexion* theory, Dr. John Aitken, in his Paper "On the Colour of the Mediterranean and other Waters" (Proc. R. S. E. vol. xi, page 472), having shown conclusively that the former theory is the correct one. "According to the *selective reflexion* theory the colour is due to the light reflected by extremely small particles of matter suspended in the water. These particles being so small they can reflect only the short waves of light, or those which belong to the blue end of the spectrum. The other theory explains the colour by supposing that water has a *selective absorption* for the rays of the red end of the spectrum—*that water is in fact a blue transparent medium.* Three different methods were adopted of testing the correctness of these rival theories, and all three proved the water of the Mediterranean to be blue by selective absorption, and show that light in passing through the water has the rays of the red end of the spectrum absorbed, and only those of the blue end transmitted."

In the same paper Dr. Aitken shows how, while the selective absorption of the water determines its colour, the brilliancy of colour is determined by the quantity of solid particles that the water carries in suspension. All the different colour phenomena of the Mediterranean are not only thus easily accounted for, but can even be imitated experimentally by means of a solution of Prussian blue and a fine white powder. "If the solution of Prussian blue is placed in a vessel, the bottom and sides of which are dark and reflect no light, then the coloured solution appears dark and colourless; but if a little of the white powder is added then the solution at once becomes brilliantly coloured. By varying the amount of the powder in the water all the varied colour effects of the Mediterranean can be reproduced, a little powder causing the solution to appear deep blue, and as more powder is added the brilliancy of the water increases, and its colour changes from blue to chalky blue-green."

The influence of the colour of the floating particles themselves on the appearance of the water is also pointed out, and the greenness of our northern seas attributed in part to the reflecting particles being yellow. Dr. Aitken's investigations included the examination of well waters, which were found to vary in colour from blue to yellowish-brown, and it was observed that the more transparent a water was, the nearer its colour was to blue. A series of experiments were also made to determine the colour of distilled water, which proved to be almost exactly that of Prussian blue. "As the amount of colour in the Mediterranean water, and in the bluish well waters, was as near as could be judged the same as in pure water, it does not seem necessary to call in the aid of impurities to account for the blue colour seen in lakes and seas, the colour being principally due to the water itself; and the different substances in solution, instead of making the water blue, tend to change its proper colour and make it green or yellow."

Prof. Threlfall, in "Nature," vol. lix, page 461, describes some observations that he made during a voyage from Sydney to Marseilles in 1898. He was able exactly to match the colour of the sea-water, as seen endwise in a tube 736 cm. (about 8 yards) long, by viewing mixtures of definite substances in a second tube 18 cm. (about 7 inches) long, placed alongside the first. The colour was seen in each case by looking through the tube against a white background. A formula is given by means of which the colour of a considerable depth of Mediterranean water can in this way easily be reproduced in a short tube. "Make up the following solution:—Water, 500 cc.; soluble Prussian blue, ·001 grm.; saturated lime water just precipitated by the smallest excess of bicarbonate of soda, 5 cc. This mixture, when viewed through a tube 18 cm. long, will show with considerable precision the colour of a sample of water from the Mediterranean, lat. 36° 27′ N; long. 17° 51′ E of Paris. By using various lengths of tubes I found that when a match has once been made, it can be preserved (within the limits tested) by increasing the amount of Prussian blue proportionally to the length of the column of water under investigation. . . . The majority of the samples of water examined by me took 25% less blue to match than the example quoted; and when the water was soiled by the tube, and perhaps at other times, it was necessary to add an amount of picric acid rising to a large proportion of the Prussian blue, and, of course, giving a green solution. The transparency of the water is estimated by the amount of precipitated chalk it is necessary to add."

With regard to the unusually green and clear water on the west and south-west coast of Australia, Prof. Threlfall suggests that the greenness may be caused by the solution of a yellow colouring matter from dead or even living seaweed.

Lord Avebury, in his work on "The Scenery of Switzerland," page 217, describes the experiments of Forel, who showed that the blue water of the Lake of Geneva could be turned into a green exactly similar to that of the Lake of Lucerne by the addition of a small quantity of water coloured yellow by the infusion of peat.

Since writing the above, my attention has been drawn to the fact that Sir Humphry Davy, in his anonymous work, "Salmonia, or Days of Fly-Fishing" (published in 1828, the year before his death) gave an account of the colour of water substantially in accordance with modern ideas. The only point in which his views seem incomplete is with regard to the luminosity of water, which as shown by Aitken, depends upon the number and colour of the suspended particles. As the book is now somewhat rare, and Davy's work on the subject—almost his latest contribution to science—not generally known, I give the complete passage here.

"POIETES.—You, Halieus, must certainly have considered the *causes* which produce the colours of waters. The streams of our own island are of a very different colour from these mountain rivers, and why should the same element or substance assume such a variety of tints?

"HALIEUS.—I certainly have often thought upon the subject, and I have made some observations and *one* experiment in relation to it. I will give you my opinion with pleasure, and as far as I know, they have not been brought forward in any of the works on the properties of water, or on its consideration as a chemical element. The purest water with which we are acquainted is undoubtedly that which falls from the atmosphere. Having touched air alone, it can contain nothing but what it gains from the atmosphere, and it is distilled without the chance of those impurities which may exist in the vessels used in an artificial operation. We cannot well examine the water precipitated from the atmosphere as rain without collecting it in vessels, and all artificial contact gives more or less of contamination; but in snow, melted by the sunbeams, that has fallen on glaciers, themselves formed from frozen snow, water may be regarded as in its state of greatest purity. Congelation expels both salts and air from water, whether existing below, or formed in, the atmosphere; and in the high and uninhabited regions of glaciers, there can scarcely be any substances to contaminate. Removed from animal and vegetable life, they are even above the mineral kingdom; and though there are instances in which the rudest kind of vegetation (forms of the fungus or mucor kind) is even found upon snows, yet this is a rare occurrence; and red snow, which is occasioned by it, is an extraordinary and not a common phenomenon towards the pole, and on the highest mountains of the globe. Having examined the water formed from melted snows on glaciers in different parts of the Alps, and having always found it of the same quality, I shall consider it as pure water, and describe its characters. Its colour, when it has any depth, or when a mass of it is seen through, is bright blue; and, according to its greater or less depth of substance, it has more or less of this colour: as its insipidity and its other physical qualities are not at this moment objects of your inquiry, I shall not dwell upon them. In general, in examining lakes and masses of water in high mountains, their colour is of the same bright azure. And Captain Parry states, that the water on the Polar ice has the like beautiful tint. When vegetables grow in lakes, the colour becomes nearer sea green, and as the quantity of impregnation from their decay increases—greener, yellowish-green, and at length, when the vegetable extract is large in quantity—as in countries where peat is found—yellow, and even brown. To mention instances, the Lake of Geneva, fed from sources (particularly the higher Rhone) formed from melting snow, is blue; and the Rhone pours from it, dyed of the deepest azure, and retains partially this colour till it is joined by the Saone, which gives to it a greener hue. The Lake of Morat, on the contrary, which is fed from a lower country, and from less pure sources, is grass green. And there is an illustrative instance in some small lakes fed from the same source, in the road from Inspruck to Stutgard, which I observed in 1815 (as well as I recollect) between Nazareit and Reiti. The highest lake fed by melted snows in March, when I saw it, was bright blue. It discharged itself by a small stream into another, into which a number of large pines had been blown by a winter storm, or fallen from some other cause: in this lake its colour was blue-green. In a third lake, in which there were not only pines and their branches, but likewise other decaying vegetable matter, it had a tint of faded grass green; and these changes had occurred in a space not much more than a mile in length. These observations I made in 1815: on returning to the same spot twelve years after, in August and September, I found the character of the lakes entirely changed. The pine wood washed into the second lake had disappeared; a large quantity of stones and gravel washed down by torrents, or detached by an avalanche, supplied their place: there was no perceptible difference of tint in the two upper lakes; but the lower one, where there was still some vegetable matter, seemed to possess a greener hue. The same principle will apply the Scotch and Irish rivers, which, when they rise or issue from pure rocky sources, are blue, or bluish green; and when fed from peat bogs, or alluvial countries, yellow, or amber-coloured, or brown—even after they have deposited a part of their impurities in great lakes. Sometimes, though rarely, mineral impregnations give colour to water: small streams are sometimes green or yellow from ferruginous depositions. Calcareous matters seldom affect their colour, but often their transparency, when deposited, as is the case with the Velino at Terni, and the Anio at Tivoli; but I doubt if pure saline matters, which are in themselves white, ever change the tint of water.

"ORNITHER.—On what then does the tint of the ocean depend, which has itself given name to a colour?

"HALIEUS.—I think probably on vegetable matter, and perhaps, partially, on two elementary principles, iodine and brome, which it certainly contains, though these are possibly the results of decayed marine vegetables. These give a yellow tint, when dissolved in minute portions in water, and this, mixed with the blue of pure water, would occasion sea green. I made many years ago, being on the *Mer de glace*, an experiment on this subject. I threw a small quantity of iodine, a substance then recently discovered,[1] into one of those deep blue basins of water, which are so frequent on that glacier, and, diffusing it as it dissolved with a stick, I saw the water change first to sea green in colour, then to grass green, and lastly to yellowish green; I do not however, give this as a proof, but only as a fact favourable to my conjecture.

"POIETES.—It appears to me to confirm your view of the subject, that snow and ice, which are merely pure crystallized water, are always blue, when seen by transmitted light. I have often admired the deep azure in crevices in masses of snow in severe winters, and the same colour in the glaciers of Switzerland, particularly at the arch where the Arve issues, in the Valley of Chamouni."

[1] Iodine was discovered by Courtois in 1811, and its elementary nature established by Davy towards the end of 1813.

C. E. Wanless.] [*Scarborough*

XXXVII.

"OFF TO SEA."

CHAPTER IV

COLOURS IN RIPPLED WATER

WE have now arrived at the last division of our subject, in which we have to take into account the colour of the water in combination with a *rippled surface*. We have seen that when—as is usually the case with rippled water (see page 29)—the line of vision strikes the near side of a wave, it is tilted upwards from its normal direction, so that where other objects had before been seen by reflexion in the still water, the eye now receives for the most part reflected skylight. And on page 47 it was pointed out how great a depth of sky affects every part of the rippled water beneath it. Thus it is evident that the apparent colour of the water must be very largely influenced by the general tone of the sky. Though a smooth sea, looked at obliquely, reflects only a low region of the heavens, a rough sea reflects all parts up to a high altitude. And, if in still water on a clear day the lower or paler sky is visible, in rippled water it will be chiefly the deeper light of the upper heaven that meets the eye, with the result that wherever the breeze catches the surface the reflected blue gains in intensity, making the rippled water look darker than the smooth. Should there be black storm-clouds overhead, the water, as soon as it is ruffled by the rising wind, reflects their gloomy colour, assuming in response a dark and threatening aspect. It sometimes happens that there is a high bank of dark clouds rising up from the horizon, and overhead a lighter region, in which case the less common effect of the rippled surface appearing lighter than the smooth may be given. Towards sunset, in the same way, the oily-looking streaks of smooth water in a very calm sea often appear yellow or orange by contrast with the more rippled—and therefore bluer—surface surrounding them, reflecting, as they do, a lower part of the sky where warmer tones predominate. Plates XXXVIII and XXXIX are photographs illustrative of this point. In the upper view, taken at the narrow mouth of a West Highland loch, the tide is flowing rapidly in a strong current from the left. The wind (also from the west) ruffles the mass of the water, but leaves the surface of the current, travelling in the same direction as itself, comparatively smooth. So that, while the greater part of the water reflects the sky, the path of the current reflects the hillside and thus becomes conspicuous as a dark stream winding its way through the lighter ripple. There is also a patch of smoother and therefore dark-looking water under the lee of the island on the left. The lower view shows an effect of the same nature, but with the opposite result. The whole surface of the water is ruffled, some parts of it more than others. In this case the smoother portion reflects chiefly light from the low bright region of the sky, whilst the rougher portion reflects the darker clouds above.

But, in addition to this effect, we have the further

XXXVIII.

KYLESKU FERRY.

Rippled water lighter than smooth.

XXXIX.

CUCHULLIN HILLS.

Rippled parts darker than smooth.

one that the *colour of the water itself becomes more apparent under the new conditions; for the line of vision now strikes the surface more directly*, (*where it hits the near sides of the waves*) *than it would if the water were still.* In our previous diagrams we have represented the surface of a wave by means of a continuous curve, which nearly resembles

the outline of a gentle ripple, but the billows of rougher water are covered with smaller waves whose crests rise up into cusps or points, thus:

so that they present surfaces nearly perpendicular to the direction of vision, which, as shown in the previous chapter, is the position most favourable for seeing the colour of water. For of course it is the same thing whether we change our position so as to look more directly at the level surface of still water, or whether the surface of the water be inclined so as to face more fully our line of vision.

The conditions are exceedingly complicated. The eye is bewildered by light reflected from a great part of the sky by countless moving facets, and the dancing waves seem to mock our feeble attempts at analysis. But it will nevertheless help us to bear this principle in mind; and knowing that the more abruptly we look at the surface of the water, the more we may expect to see of its true colour, we shall be the better able to distinguish this colour wherever it appears from that due to reflexion. Anyone who has stood on the sea beach is aware that in the curl of a wave breaking towards him the colour of the water is plainly visible, and in the same way we see it, though in a less degree in every wave surface turned towards us. The stronger the ripple, the more steeply the little waves stand up to face us, and so the more noticeable this colour becomes.[1] Against this, however, must be set the consideration that the effect referred to at the beginning of this chapter, namely, that the rippled parts reflect to the eye chiefly light from the higher, and therefore probably darker, regions of the sky, will, of course, unless the sky be uniform in tone (which is seldom the case) somewhat obscure the perception of this further effect.

The waters of our English lakes and rivers are comparatively clear and colourless, and it is therefore only when they are in a turbid state that they offer an illustration of this point; with the sea, on many parts of the coast, it is otherwise; there the water generally carries enough solid matter to reveal its colour, and thus, in sunshine, at any rate, the result we are considering is conspicuously produced. This is perhaps most readily observed when adjacent portions of the sea present different colours. The

[1] It will be remembered that in the experiments described on page 75 we found that the blue water, when looked at very obliquely, seemed practically colourless. If we now take the same basin of blue water and placing the eye in the same position (as shown in Fig. 27) stir the water, so as to give movement to the surface, we shall find that its colour at once appears on the near sides of the little waves. Though we are looking obliquely at the surface of the water as a whole, each ripple presents a surface inclined at a considerable angle to the direction of vision, and thus reveals its colour, just as the whole body of still water showed its colour to one standing over it and looking abruptly at its level surface.

colour of the sea near the shore, is, as we have said, green, and in one spot, where the water is in movement, the ripples on the surface show, as has been explained, this green to the eye. Close by there may be a portion of still water, sheltered by a ledge of rocks, and in this the smooth surface, reflecting the blue sky and showing little or no local colour, serves to emphasize the greenish tint which the neighbouring water, in virtue of its ripples, is displaying. A violent gust of wind, suddenly striking a nearly calm surface, brings out the colour of the water to a still greater extent, owing to the greater steepness of the minute waves to which it gives rise.

When little waves are lapping on the smooth sand, it is the colour of the sand, rather than of the water, too shallow to show any colour of its own, that we see in the upturned surfaces. Here the difference of colour between the near and far sides of the waves is very plainly seen, the latter showing nothing of the colour of the sand beneath them, but only reflected blue from the sky. In shallow estuaries a delightful variety of colour is caused in this way in sunny weather. In the strongly rippled surfaces, wherever the water is turbid or very shallow, it shows the pink colour of the sand, and where it is deeper and clearer (but still strongly rippled) it looks a bright green; whilst the smoother parts of the water, seen obliquely, show only sky reflexion. We have here a notable illustration of the principle we are discussing, viz., the increased power of seeing *into* water that we obtain by looking more abruptly at its surface. When a stiff breeze is blowing straight into a shallow sandy bay, as often happens on the west coast of Scotland, the contrast is still more striking. From a little distance the turbid water looks the colour of sand, whilst the margin of smooth wet sand along the shore takes that of the sky.

It is thus clear that, taking the waves *individually*, their nearer sides show more local colour and their farther sides more reflected colour. And when there is no breeze to roughen the surface, more local colour and less sky reflexion may usually be detected looking across the wave crests than looking along them. But in a general effect the local colour becomes most conspicuous when a stiff breeze is blowing, so that the whole surface of the water is ruffled, especially if one is looking into the wind, for then the line of vision strikes abruptly the steep sides of the advancing waves and their farther sides are out of sight. Thus the local colour is very prominent, whilst the reflected colour is reduced to a minimum. The colour of the Alpine lakes already referred to appears for the same reason most brilliant on a windy day, particularly if the sun is in an opposite quarter and shining directly onto the farther sides of the waves, the edges of which then show glimpses of their own pure emerald. The bright green colour of the glacier-fed lakes, seen under these conditions, mingled with the deep blue reflected from the sky, forms a combination scarcely equalled in brilliancy by anything in nature.

But the local colour is often entirely masked by the brightness of the reflexion on the rippled surface. For instance, as we stand by the shore of one of these mountain lakes, with its natural colour of undoubted brilliancy, and watch the breeze rippling the water in streaks or patches a few yards from us, and thus disturbing the image of the mountain on the opposite shore, we look in vain in the ripples for the added local colour, which we might have expected, and can detect nothing but a gleam from the sky above. Follow one of these streaks sideways, however, into the reflexions of the tall trees that line the bank close at hand, where the upturned faces of the ripples reflect the tree-tops and not the sky (so that there is no interruption of the dark image by bright sky light) and there the colour of the water itself, blue or green, as the case may be, can again be detected, affording an illustration of the fact that (as pointed out on page 78) the local colour of water reveals itself most in the reflexion of dark objects. In Plate XXXIX, if the water were not too clear, its colour would in all probability appear in the more ruffled parts. But in Plate XXXVIII we could hardly expect the rippled water, which reflects the sky, to show more local colour than the smooth water, which reflects the dark hillside. If, however, the hill were high enough, or near enough, to cast an interrupted reflexion over the whole water, rippled and smooth, then again the local colour might be discerned in the rippled surface. These instances show how, under the complex conditions at which we have arrived in this final stage of our inquiry, it is impossible to give to the artist hard and fast directions, one effect often overpowering the other to such an extent that the latter is hardly perceptible.

* * * * *

It would be outside the scope of this essay to attempt to discuss fully and systematically the varied and complex colour effects seen at sea, but, beyond the general outline which we have presented, a few remarks on some special features will not be out of place.

The waters of the Mediterranean are rich in colouring, and will furnish illustration of certain points. At some distance from land, the churning of the screw or the foam of the breakers shows an almost pure blue, but nearer shore (probably owing to contamination of some kind) the water assumes a greenish tinge, which becomes still more marked in the harbours. This brilliant colour, it need hardly be said, is very different from the chalky green of the Straits of Dover that often greets us on our return to English waters from sunnier seas. Looking down into the *clear* depths of the Mediterranean, however, the water often seems to be of a slightly purplish or violet blue, and does not show any sign of green. Even on a dull day, when looked at perpendicularly from the deck or in the near sides of the waves, it still appears of this beautiful deep ultramarine,—so that the colour is evidently not due to reflexion from the blue sky,—whilst the farther sides of the waves, being tilted away from us, reflect strongly the gray light from the clouds. The most probable explanation of this difference is that sea water, in common with many other liquids, possesses the property of showing a somewhat different colour according as one is looking through a thick or a thin layer of it.[1] In this case the particles in the

[1] A solution of Prussian blue, with which the sea water was matched in Threlfall's experiments, described on page 84, has this

depths give a thick layer, whilst the air bubbles in the wake, being near the surface, furnish a thin layer of water; the former appears violet-blue or ultramarine, the latter greenish or cyan-blue.

When at anchor in full sunshine, the colour of the water is most vividly seen towards midday if the observer looks over the rail at the sunlit surface close to his own shadow, for in this position he is looking straight at the illuminated sides of the floating particles (as we do at the illuminated surface of the moon when at the full), and they therefore reflect more light to the eye than if the direction of vision differed considerably from that of illumination. Here the colour will be more brilliant than in the shadow of the vessel, where (as explained in the last chapter, page 77) more of the sky colour and less local colour is seen. Looking over the other side of the boat, the colour will not show up so strongly—unless in the dark reflexion of a boat lying alongside, where, the sky reflexion being interrupted, the colour of the water is again very obvious.

By the shore there will often be seen a patch of bright green, where a shelving rock or pebbly beach

property. If two bottles of colourless glass, one large and the other small, be filled from the same solution of Prussian blue, and placed against the same white background, the difference, not merely in shade, but actually in colour, between a thin and a thick layer of the solution will be seen at once. Indeed, one bottle is sufficient to show it, the colour varying with the direction in which one looks through the bottle, or with the strength of the transmitted light. The paler parts are distinctly green, the darker show no signs of it; and in the deepest parts, as when looking down on to the shoulder of the bottle, the violet tinge appears.

projects into the shallow water. The brilliant colour here may be partly caused by the yellow or orange colour of the stone or shingle. A bank of pink sand seen through the blue water sometimes shows as a purplish streak, whilst the margin of deeper purple at the water's edge is due to masses of seaweed and ink-like stains on the submerged rocks. These pink or purple-blue patches on the water can be seen almost anywhere along our coasts. That rocks, which at low tide are seen to be of a dark purple colour, or even the darker kinds of seaweed, should give rise to such patches seems natural enough, but it may occasion some surprise to find that they are often caused by orange or olive-green seaweed. So different is the natural colour of the seaweed to the delicate "pink madder" colour that it assumes under water, that the casual observer will hardly connect the two, until he has satisfied himself on this point by closer inspection (see page 110).

In very calm waters, dark lines, having almost the appearance of shadows, are often seen stretching across the surface in the distance; these will prove on approach to consist of minute ripples reflecting light from the upper sky. The water is seldom smooth enough to reflect the lowest region of the sky. Thus it is that the golden streak of light from the setting sun disappears before the sun touches the horizon. But if the sea is absolutely calm, the horizon line may entirely disappear—an effect said to be of common occurrence in the Mediterranean—so exactly is the colour of the sky repeated in the water. For we owe our usual horizon line to the fact that the ripples on the water reflect the higher part of the sky, which being darker, marks a distinction between the water and the lighter sky immediately above it. But should the disposition of clouds happens to be such that the lowest part of the sky is darker than that above, then the distant water will in all probability appear lighter than the dark clouds beyond.

Beautiful results are produced by the harmonious blending of colours in rippled water, as each face of the ripple catches a different tone. When seen from a little distance, these tints are completely mingled in the eye, giving the sensation of a new colour. Thus, the reflexion of barren red rocks in rippling water, which reflects also the blue sky, gives a purple formed by the mixture of these two colours. The lovely tints seen in rippled water at sunset are probably due in some degree to this combination of lights from differently coloured portions of the sky (see page 106). Near at hand, where the ripples can be seen individually, more than one reflected colour may be distinguished in each, and in addition to this at times the local colour on its nearer side. Thus it is sometimes possible to make out three, or even four, distinct colours on each little wave. The combination of white cliffs under a slope of green grass with a blue sky above gives a simple instance of three definite tones. The perpetual motion of the water, though so rapid as to make it very difficult to discern clearly the different tones, is yet not rapid enough to make them appear to be blended into one colour, so that the effect has to be imitated—at least in the foreground of a picture—by small patches of colour. It is hardly necessary to point out that the mere mechanical mixture of these colours on the palette would produce a totally different hue to that given by their true mixture, such as the general tone of the rippled water, as seen from a little distance, caused by the union of the several colours reflected.[1] But the true resultant tone can be given in the same way by viewing a

[1] On page 66 we referred to the absorption of rays of certain definite wave-lengths in passing through coloured substances, such as red glass. In the same way light passing through two glasses of different colour undergoes two absorptions. If we look through a blue and yellow glass placed one over the other we see green. The blue glass, besides transmitting blue light, allows a certain amount of green light to pass; the yellow glass also transmits some green light; but the blue rays are cut off by the yellow glass and the yellow rays by the blue glass. It is thus only the green rays which are able to escape the twofold absorption and so reach the eye. Similarly with a mixture of blue and yellow pigments the resulting colour is green; the greater proportion of the incident rays penetrates to a small depth below the surface before being reflected, and of these rays all except the green are absorbed by the combined action of the blue and yellow particles of which the mixture is composed.

But blue light and yellow light when mingled do not produce green, but *white*. This can be shown by throwing beams of yellow and blue light on to the same white surface in a darkened room, or more simply by spinning a disc coloured partly yellow and partly blue. "Take two of Maxwell's cardboard discs, each with a radial slit, one of the discs being painted with ultramarine and the other with pale (not orange) chrome yellow. Adjust the discs so that half of one disc is concealed behind the other—as shown in Fig. 28, where *a—a* indicates the radius at which the blue disc passes under the yellow disc, and *b—b* the radius where the yellow disc passes under the blue—on rotating the compound disc, so that the eye

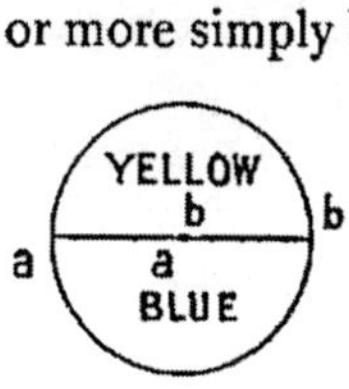

Fig. 28.

surface entirely covered with minute spots of the constituent colours at a distance sufficient to cause them to blend completely in the eye. This system of mixing coloured lights on the canvas has advantages which would seem to recommend it for the painting of the whole surface of the rippled water. It not only gives increased luminosity, but also, as pointed out by Rood, adds a lustre and transparency not otherwise obtainable. "If the coloured lines or dots are quite distant from the eye, the mixture is of course perfect and presents nothing remarkable in its appearance; but before this distance is reached there is a stage in which the colours are blended,

shall receive simultaneously blue and yellow light, not the slightest approach to green is produced, but a grey, slightly tinged with yellow. By carefully reducing and adjusting the portion of the yellow disc exposed it is possible to get rid of this yellowness and to obtain an absolutely neutral grey . . . The light reflected from a blue pigment, mingled with the light reflected from a yellow, invariably produces white of small brightness, that is, a neutral grey" ("Colour," by Prof. A. H. Church, Cassell and Co., 1887, page 80).

According to the Young-Helmholtz theory of colour-perception, now generally accepted, there are three primary colour sensations, viz., red, green and blue (or violet). The secondary colour-sensations arise from mixtures of two of these primaries; thus a mixture of red and green lights gives the sensation of yellow, green and blue together give that of sea green, and red and blue that of purple. A mixture of all three primaries in the right proportions gives white. The complementary of any given colour is that colour which must be added to it in order to produce the sensation of white. Thus yellow is the complementary of blue, purple of green, and sea green of red. For a complete account of Young's theory and its developments, see Church's manual on "Colour," from which is taken the passage quoted above.

though somewhat imperfectly, so that the surface seems to flicker or glimmer—an effect that no doubt arises from a faint perception from time to time of its constituents. This communicates a soft and peculiar brilliancy to the surface, and gives it a certain percentage of transparency; we seem to see into it and below it."[1]

Though less rich in colouring than many southern waters, our home seas can still boast a great variety of tint and tone. After rough weather they generally hold a quantity of fine sand in suspension, which lessens as a rule as the distance from the shore increases—though it is not always the nearest water that carries the most sand—so that we get gradations from a sandy yellow or pinkish tinge through blue-green to the almost pure blue that marks the deep water, where there are fewer particles, and possibly also less stain from organic matter. The change of colour may be so gradual as to be almost imperceptible, or, owing to currents carrying more or less sediment than the surrounding water, comparatively sudden. When, however, the cliffs are of granite or hard igneous rock, the sea is very clear, the proportion of floating particles being much less than along a softer coast of sandstone, chalk or clay, and consequently in deep water or over a dark bottom its colour hardly appears. But where the bottom is of bright sand, as on many parts of the coast of Cornwall, the colour of the shallow water is seen to great

[1] "Colour," by Prof. Ogden Rood (Kegan Paul and Co., 1890), page 280. Chapter XVI, from which the above quotation is taken, cannot fail to be of interest and value to the art student.

advantage. Indeed, the Cornish seas almost rival those of the Mediterranean in brilliancy, and are noted for their greener hue. The water is unusually clear, the headlands being of very hard rock separated by shallow sandy bays. This sand is largely composed of broken pieces of pink and yellow shells mixed with shining grains of quartz and mica. Its general colour is orange, and that no doubt, among other causes, influences the apparent colour of the water, making it look greener than it would over a white bottom. The skies of Cornwall are, of course, less blue than those of the Mediterranean, being usually in fine weather light and clear, and it is probably for this reason that the sea does not get so dark towards the horizon, as it does under a southern sky.

On other parts of our coast the sea depends for beauty of effect more particularly upon atmospheric conditions of colour and lighting. With a uniformly dull gray sky the water naturally looks dull also, and we have pointed out in Chapter III (page 68) how the local colour is emphasized by direct sunshine. But it is not necessary that the sun should be actually shining on the water in order that its characteristic green should show up strongly. There is often, especially in showery weather, though the sun itself is hidden, a very bright region of the sky on one side, whilst in another direction a dark, unbroken cloud-mass rises from the horizon, extending high up overhead, so that very little reflected light from that quarter reaches the eye. In this latter direction the colour of the water will be unusually conspicuous, though under the lighter part of the sky, where the reflected light is far stronger, it may be hardly noticeable. As explained on page 78, it is in the reflexion of *dark* objects that the colour of water shows most plainly. This local colour will, of course, be still more apparent when the sunlight is no longer hidden, and bursts out, striking the water under the dark clouds. Here we have a remarkable instance of dissimilarity between sea and sky. The clouds may be inky black, yet the water, lighted up by the sun from behind the spectator, appears to him of a vivid green colour.

Heavy clouds, floating in a luminous sky, or bright patches in a dull sky, give rise on the surface of the water to masses of light and shade, which take the form in the picture of vertical bands (see page 47). The effect in a lively sea may be so broad as easily to escape observation, but where there is any considerable variety of tone in the sky it is none the less existent. The darker bands in the water will of course show more local colour, and the lighter bands more reflexion colour, a very strong sky reflexion entirely masking the colour of the water.

But, in addition to these cloud reflexions, we often see definite *cloud shadows* floating across the surface of the water, and adding greatly to its beauty of colour. They are generally of a slightly purple or pinkish hue, and vary somewhat in colour with the colour of the water on which they fall (see Note at end of chapter). In the distance they assume the form of horizontal streaks or lines. Though of everyday occurrence in fine weather off the chalky shores of the Channel (see page 77), they are rarely to be seen in the clearer waters, such as those of the west Highland coast, unless indeed they happen to cross the brilliant pathway beneath the sun, when they suddenly become conspicuous in virtue of their interruption of its dazzling reflexion. The characteristic "half-shadows," so often seen in dull weather, are formed in this way by light falling on to the water through narrow openings in a lower layer of clouds from high luminous clouds above them. Sometimes also, while no direct sunshine reaches the surface, such a narrow opening allows light from some blue sky directly above it to fall on the water, with the result that we get a patch of blue amongst the gray.

We have now perhaps said enough to show that our seas, and particularly in fine weather, cannot be accused of monotony. Currents carrying more sediment than the surrounding water, varying breezes rippling one part more than another, broad masses of light and shade in the sky affecting differently the different parts of the sea beneath them by their vague, drawn-out reflexions, and the purple shadows cast by passing clouds; all these combine to present to one standing on a height and looking down at the water a charming diversity of colour and tone.

There are, however, many beautiful effects that are entirely due to reflexion. Sometimes above the crimson and gold of a sunset sky and beneath the higher blue there is a comparatively small expanse of green, which seems singled out for reflexion by the water. In a rougher sea we might get a rosy tinge from pink clouds, if there happened to be any higher up, or, with a clear sky overhead, a complete absence of such warmer colour on the water; but in the case in point we suppose a nearly calm sea. The more brilliant red and gold are too low down to be repeated (compare the disappearance of the golden streak before the sun reaches the horizon, page 38), and thus the beautiful green light floods the whole surface of the water, rendered perhaps more delicate still by blue reflected in any rippling surface from higher parts of the sky. The near sides of the waves breaking on the shore are dark, and naturally seem to assume the purple tinge, which is complementary to the green, or, if the light on the water is orange, a deep blue. At the same time, the wet sand, sloping gently down to the water, may reflect the gorgeous red of the lower sky.

The narrow bays and sea lochs of the West coast of Scotland, sheltered at their heads and stretching out to the more open sea, afford constant illustration of the difference in colour between smooth and rough water, referred to at the beginning of this chapter, owing to their reflecting light from different parts of the sky. The contrast is most noticeable towards sunset, and whereas, if the whole surface of the water were absolutely smooth, we should see the sky colours reversed in the water, we may now have them repeated—with omissions—in the same order, the distant rougher water reflecting the blue of the upper sky, and the near smoother water the warmer tints below. If these latter are very brilliant, the tone of the ruffled water often seems strangely cold by contrast with their glowing reds and yellows, and its peculiar neutral gray may be due to the mingling of almost complementary hues reflected by the ripples from differently coloured regions of the sky.

NOTE ON THE COLOUR OF SHADOWS ON WATER.

IN the last chapter it was shown how in shadow the reflexions of objects are stronger than in sunlight, but this consideration alone does not seem sufficient to account for the colours often observed. For instance, the shadow of a cloud on the brilliant blue-green waters of an Alpine lake as seen from a height appears of a distinctly pinkish or purple tinge, and the same is true to a less extent of the shadows of clouds seen at sea.

Most people, unless gifted with a very keen sense of colour, find it difficult to define at all the colour of a shadow, and indeed it is only when the eye has been trained by repeated attempts at sketching that it begins to value at all correctly the subtle gradations of tones in light and shade. Moreover, the colour of objects is changed for the beholder according to the conditions under which at the moment he sees them. Thus the patches of wet sand left by the ebbing tide, though reflecting the pure blue of the sky, sometimes appear greenish by contrast with the pink expanse of dry sand surrounding them. Such an effect, which in nature is produced by the mere juxtaposition of different tones, does not, however arise from the juxtaposition of the lifeless pigments at our disposal. These are not luminous enough to create it, and thus, to reproduce it on the canvas, the artist has, perhaps unconsciously, to exaggerate the natural intervals of colour. We have said that it is generally difficult to define the colour of a shadow, but there is one case in which it is comparatively easy, namely, when the ground is entirely covered with snow, so that all disturbing colours are removed. On page 9 we referred to the irregular reflexion from the surface of rough or unpolished objects, and pointed out how in looking at such an object the eye receives light from all sides by reflexion in the countless small surfaces that go to make up the whole. It is evident therefore that the apparent colour of all such unpolished objects will be affected by the colours of surrounding objects—a fact with which all artists are familiar. This modification of the normal colour of an object by its surroundings is most readily observed in the case of a white surface, such as a piece of white drapery or a whitewashed wall, which in the neighbourhood of well-illuminated coloured objects, may show a wonderful variety of hues and shades. In the case of the shadows on the snow we have practically only two tones to deal with—the pure white of the snow in sunshine, and the deep blue reflected from the sky above—so that the conditions are as simple as possible. Such blue shadows, seen at their brightest in the clear atmosphere of high altitudes, are well known to visitors to winter Alpine resorts. Now the shadow on the snow of a colourless object, such as a wooden post, appears blue, because the direct white light from the sun is cut off by the post and the shaded part receives only light from the sky above it, consisting largely of blue rays. For general purposes it is near enough to say that the shadow is blue, though as a matter of fact it appears slightly more violet in hue than the sky with which one compares it.

This purplish or violet tone is probably due to the fact that in the shadows is seen chiefly light reflected from the higher parts of the sky, which contains a larger proportion of violet rays than that from regions near the horizon.[1] It

[1] Dr. Shelford Bidwell has observed that the "after-image" of a bright white object, when projected upon a gray background, appears under certain conditions to be purple and not merely a darker gray, which phenomenon he has attributed to the fact (pointed out by Rood) that the green colour-sensation is more readily fatigued than the others. The disproportionate weakening of the green sensation would make a physically gray object appear purple, for the red and blue-violet elements would predominate. The purplish ele-

is a matter of common observation that this latter part of the sky appears to be of a purer blue than the higher regions overhead, which have a decidedly violet tinge. Now it is naturally with the blue of the lower sky that we compare the shadow, and by contrast it appears somewhat purple. This may be verified by a simple experiment. We can take a small mirror and lay it flat on the snow in the shadow of some post or other convenient object. Looking straight down into the mirror we find the deep colour reflected from the overhead sky so much stronger than that of the surrounding shadow that it is difficult to compare the two tones; but stepping back and again looking into the mirror, we naturally see a lower part of the sky reflected in it, and we now observe that the colour shown in the mirror is distinctly bluer than that of the shadow. This will be most clearly noticed if we draw back to such a position that, as we look obliquely at the mirror, the tone of its blue reflexion is of the same depth as the tone of the more purple colour of the shadow. The mirror reflects light *regularly* from a low and comparatively pure blue part of the sky, whilst the shaded snow, owing to its innumerable facets turned in all directions, reflects light *irregularly* from all parts of the sky and very largely from those dark and more violet regions overhead that do not enter into the field of vision.

Thus the shadow of an object in bright sunshine almost always receives a large proportion of blue or violet-blue light, and in this way we can explain the "cool" tones of shadows under more ordinary conditions. It is often said that "shadow is cooler than shade," by which is meant that the shadow cast by an object is cooler in tone than the shaded side of that object. This is generally the case and is due to the fact that the "shade" receives more warm light from surrounding objects than the surface covered by the

ment in these shadows may therefore, as suggested by Prof. Threlfall, be partly due to physiological causes depending on a fatigue of the eye from looking at bright snow.

shadow, which, in all probability, faces directly towards the blue sky. For instance, the shadow of a red-brick chimney falls upon a red-tiled roof. The shady side of the chimney reflects a certain amount of light from the blue sky, but also a good deal from the warm red tiles of the surrounding roof, whilst the shaded portion of the roof, being turned towards the sky, reflects little but blue light and thus appears cooler in tone.

In the case of the cloud shadow on the lake, however, something more than the mere blue or violet-blue colour of the light falling on the shaded surface is needed to account for the purplish tone of the shadow. The shadow may still be pink or purple, even when the greater part of the sky is covered with white clouds. But in shade there is less light reflected from floating particles than in sunshine, so that the shadow hides the colour of the water and at the same time renders the surface reflexion (from blue sky or white clouds, as the case may be) more prominent; and we may account for the apparent colour of the shadow on the supposition that the surface-reflected light there tends to assume by contrast the red or purple which is complementary to the blue-green of the surrounding water in sunshine.

This is merely thrown out as a suggestion. It is an undoubted fact that when two different colours are brought into contact their apparent difference is increased, each seeming tinged with the complementary colour of the other. A white or neutral gray patch on a bright green ground looks purplish, a blue patch on green tends to violet. And it has been noticed that, in the case of cloud shadows on water, the greener the water the pinker the shadow appears to be.

A similar question arises with regard to the pink patches caused by submerged seaweed referred to on page 98. Are we to ascribe the colour entirely to the effect of the double absorption of light—in its passage (i) through the water, and (ii) through the surface layer of the seaweed—so that the rays which finally emerge from the water give the sensation of purple? The fact that the resulting colour appears to be independent of the colour of the seaweed and to arise from *any* dark bottom, would point to its being also an effect of contrast. Moreover the pinkish colour seems to be more noticeable when looking obliquely, so that one is inclined to put it down, as in the case of the cloud shadows, to the surface-reflected light becoming tinged with the complementary of the surrounding colour. But there is much to be said on both sides of this difficult and complicated question.

INDEX